THIS PROJECT INCLUDES THESE PARTS.

ⓡ HOOKUP WIRE (20 AWG)
#278-1222

MILK JUG TOP

FOAM TAPE, DOUBLE-SIDED

GLUE

ⓡ 9V RECORDING MODULE
#276-1323

ⓡ 9V BATTERY
#23-866

ZIP TIES

FIND MORE WEEKEND PROJECTS AT **MAKEZINE.COM/WEEKENDPROJECTS**

brass tab. The weight of the object keeps the clothespin open and armed. Camouflage the booby trap by placing something else in front. If anyone lifts the book — whoop! whoop! — the alarm goes off. The snoop is busted!

» Hide it behind your lunch inside the fridge at work. If anybody touches it, the booby trap blasts: "Keep your hands off my lunch!"

» Tie a string to a small piece of paper and slip it between the clothespin jaws. Then tie the other end to any pilferable object. Use thin monofila-ment fishing line for a nearly invisible alarm.

» Open the jaws and slip the trap under a closed door. If anybody opens the door, you'll know it. "Hey, you kids — get back in bed!" You can also arm a drawer or sliding door.

» Leave a friendly audio reminder for some-one special. Arm their cellphone or car keys and they'll hear you say: "Don't forget — romantic dinner tonight!" They'll really get the message from the spy who loves them!

— *Bob Knetzger*

Make: Volume 35

34
MAKER FOR LIFE!

38

FIERY FAMILY: Jon Sarriugarte, Kyrsten Mate, and daughter Zolie with one of their Serpent Twins. Photograph by Gregory Hayes. Art direction by Jason Babler.

Vol. 35, July 2013. MAKE (ISSN 1556-2336) is published quarterly by Maker Media, Inc. in the months of January, April, July, and October. Maker Media is located at 1005 Gravenstein Hwy. North, Sebastopol, CA 95472, (707) 827-7000. SUBSCRIPTIONS: Send all subscription requests to MAKE, P.O. Box 17046, North Hollywood, CA 91615-9588 or subscribe online at makezine.com/offer or via phone at (866) 289-8847 (U.S. and Canada); all other countries call (818) 487-2037. Subscriptions are available for $34.95 for 1 year (4 quarterly issues) in the United States; in Canada: $39.95 USD; all other countries: $49.95 USD. Periodicals Postage Paid at Sebastopol, CA, and at additional mailing offices. POSTMASTER: Send address changes to MAKE, P.O. Box 17046, North Hollywood, CA 91615-9588. Canada Post Publications Mail Agreement Number 41129568. CANADA POSTMASTER: Send address changes to: Maker Media, PO Box 456, Niagara Falls, ON L2E 6V2

14

THE TASTE OF TOMORROW TODAY.

REAL COLA TASTE.
60% LESS SUGAR.*

DRINK IT
TO BELIEVE IT.®

*Than PEPSI-COLA. PEPSI-COLA, PEPSI NEXT, the Pepsi Globe and
DRINK IT TO BELIEVE IT are registered trademarks of PepsiCo, Inc.

Make: Volume 35

READ ME Always check makezine.com/35 before you get started on projects. There may be important updates or corrections.

⌃ **TINT MAKER:**
Household materials + electrolysis = metal oxide.

140

⌃ **Keepon Hacking:**
Mod a My Keepon toy to be teleoperated by connecting an Arduino to it.

124

PUT THIS IN YOUR POCKET!

129
TRONTASTIC:
Use EL tape and velcro to approximate Jeff Bridges' glowing druid look from the film.

> "The dangers of life are infinite, and among them is safety."
> —Johann Wolfgang von Goethe

FOUNDER & PUBLISHER
Dale Dougherty
dale@makezine.com

EDITORIAL DIRECTOR
Ken Denmead
kdenmead@makezine.com

VICE PRESIDENT
Sherry Huss
sherry@makezine.com

EDITORIAL

EDITOR-IN-CHIEF
Mark Frauenfelder
markf@makezine.com

MANAGING EDITOR
Cindy Lum
clum@makezine.com

PROJECTS EDITOR
Keith Hammond
khammond@makezine.com

SENIOR EDITOR
Goli Mohammadi
goli@makezine.com

SENIOR EDITOR
Stett Holbrook
sholbrook@makezine.com

TECHNICAL EDITOR
Sean Michael Ragan
sragan@makezine.com

ASSISTANT EDITOR
Laura Cochrane

STAFF EDITOR
Arwen O'Reilly Griffith

COPY EDITOR
Laurie Barton

EDITORS AT LARGE
Phillip Torrone
David Pescovitz

DESIGN, PHOTOGRAPHY & VIDEO

CREATIVE DIRECTOR
Jason Babler
jbabler@makezine.com

SENIOR DESIGNER
Juliann Brown

DESIGNER
James Burke

ASSOCIATE PHOTO EDITOR
Gregory Hayes
ghayes@makezine.com

VIDEOGRAPHER
Nat Wilson-Heckathorn

WEBSITE

WEB DEVELOPER
Jake Spurlock
jspurlock@makezine.com

WEB DEVELOPER
Cole Geissinger

WEB PRODUCER
Bill Olson

MAKER FAIRE

PRODUCER
Louise Glasgow

MARKETING & PR
Bridgette Vanderlaan

PROGRAM DIRECTOR
Sabrina Merlo

SPONSOR RELATIONS COORDINATOR
Miranda Mager

SALES & ADVERTISING

SENIOR SALES MANAGER
Katie Dougherty Kunde
katie@makezine.com

SALES MANAGER
Cecily Benzon
cbenzon@makezine.com

SALES MANAGER
Brigitte Kunde
brigitte@makezine.com

CLIENT SERVICES MANAGER
Sheena Stevens
sheena@makezine.com

CLIENT SERVICES MANAGER
Mara Lincoln

MARKETING

SENIOR DIRECTOR OF MARKETING
Vickie Welch
vwelch@makezine.com

MARKETING COORDINATOR
Meg Mason

MARKETING COORDINATOR
Karlee Vincent

MARKETING ASSISTANT
Courtney Lentz

PUBLISHING & PRODUCT DEVELOPMENT

CONTENT DIRECTOR
Melissa Morgan
melissa@makezine.com

DIRECTOR, RETAIL MARKETING & OPERATIONS
Heather Harmon Cochran
heatherh@makezine.com

VICE PRESIDENT OF COMMERCE
David Watta

OPERATIONS MANAGER
Rob Bullington

CHANNEL MANAGER
Kaitlyn Amundsen

PRODUCT DEVELOPMENT ENGINEER
Eric Weinhoffer

MAKER SHED EVANGELIST
Michael Castor

COMMUNITY MANAGER
John Baichtal

EXECUTIVE ASSISTANT
Suzanne Huston

FABRICATOR
Daniel Spangler

PUBLISHED BY
MAKER MEDIA, INC.
Dale Dougherty, CEO

Copyright © 2013
Maker Media, Inc.
All rights reserved.
Reproduction without
permission is prohibited.
Printed in the USA by
Schumann Printers, Inc.
Visit us online:
makezine.com
Comments may be sent to:
editor@makezine.com

CUSTOMER SERVICE
cs@readerservices.
makezine.com

Manage your account online, including change of address:
makezine.com/account
866-289-8847 toll-free
in U.S. and Canada
818-487-2037,
5 a.m.–5 p.m., PST
Follow us on Twitter:
@make   @makerfaire
@craft   @makershed
On Google+: google.com/+make
On Facebook: makemagazine

CONTRIBUTING EDITORS
William Gurstelle, Brian Jepson, Charles Platt, Matt Richardson

CONTRIBUTING WRITERS
Massimo Banzi, Colin Berry, Josh Burker, Christopher Dodsworth,
Ian Gonsher, Matt Griffin, Saul Griffith, Steve Hoefer, Brad Huffman,
Bob Knetzger, Sean McBeth, Marek Michalowski,
Forrest M. Mims III, Craig Newswanger, Dan Pollino, Tom Rodgers,
Rick Schertle, Kevin Simon, Eric Smillie, Jonathan Thompson,
Scott W. Vincent, Marc de Vinck, Michael Wernecke

CONTRIBUTING ARTISTS:
Shane Deruise, Nick Dragotta, Nate Van Dyke, Natalie Jeday,
Rob Nance, Damien Scogin, Julie West, Shannon Wheeler

ONLINE CONTRIBUTORS
John Baichtal, Kipp Bradford, Meg Allan Cole, Michael Colombo,
Jimmy DiResta, Nick Normal, Haley Pierson-Cox,
Andrew Salomone, Karen Tanenbaum, Glen Whitney

TECHNICAL ADVISORY BOARD
Kipp Bradford, Evil Mad Scientist Laboratories, Limor Fried,
Saul Griffith, Bunnie Huang, Tom Igoe, Steve Lodefink,
Erica Sadun, Marc de Vinck

INTERNS
Kelley Benck (engr.), Uyen Cao (ecomm.), Eric Chu (engr.), Craig Couden
(edit.), Paloma Fautley (engr.), Sam Freeman (engr.), Gunther Kirsch
(photo), Raghid Mardini (engr.), Brian Melani (engr.), Nick Parks (engr.)

PRINTED WITH SOY INK®

CONTRIBUTORS

Shannon Wheeler (*Bringing Makey the Robot to Life*) is the Eisner Award–winning creator of the comic and opera *Too Much Coffee Man*. Wheeler has contributed to a variety of publications including *The New Yorker* and *The Onion*. He currently lives in Portland, Ore. with his cats, chickens, bees, girlfriend, and children. He just built a couple rooms in his basement and he's about to build a chicken coop and run. Wheeler has multiple books that are pretty easy to find. His weekly comic strip is published in various alternative weeklies and online at tmcm.com.

Ian Gonsher (*Coffee Shop Construction Toy*) is an artist and designer living in Providence, R.I., where he teaches in the School of Engineering at Brown University. In his classes, students explore various strategies for creative thinking and making. He is particularly interested in how creative processes play out across disciplines and in the development of strategies for recognizing and generating abundant value through exploratory play. This often involves working on collaborative projects with students, some of which can be seen at browncreativemind.com.

Nick Parks (MAKE engineering intern) is a college student studying mechanical engineering. He likes to build, destroy, and build things again. When he isn't making things, you can find him driving remote-controlled vehicles or riding his handmade longboard down the steep and adventurous hills of Sebastopol. Parks' life goal is to eventually start his own company to create products that will help people to do their best to conserve resources and preserve the great features of the planet.

Craig Newswanger (*The Six-Pack Tesla Coil*) of Austin, Texas, is a holographer, inventor, artist, and maker. Craig's varied interests include: optics, electronics, photography, holography, astronomy, computer music, computer graphics, mechanics, and the history of science. He has dreamed of building a workshop where he could build stuff like telescopes, Tesla coils, kites, and anything else he can think of. He really likes to build things and make them work! He works for Zebra Imaging, builds things at the Resonance Studio Workshop, and is also a behind-the-scenes member of ArcAttack.

Chris Dodsworth (*Raygun Vector Weapon*) is a joinery manufacturer from Lincolnshire, in the U.K., who started designing analog audio circuits back in 2009 under the name Symetricolour. Inspired by a love of synth equipment but wanting to create more experimental sounds, he created electronic designs and kits of small lo-fi devices that produced everything from soothing beeps to raging torrents of noise. More recently he has been using these devices, such as the Noise Hero and the Beep Poet 2, while playing keyboards in the band Robots Found Errors, gigging around Lincolnshire.

Shane Deruise (*Pyro*) was born in North Carolina and grew up next to the sea. The eldest of two children, he displayed no interest in becoming a fisherman, and after high school he decided to travel the country, not with an amazing circus, but as a professional wrestler. Deruise loves to shoot bands, and portraits of people who have never been in magazines. He dreams of traveling the world with his beautiful wife and amazing son, Bresson, who is the self-proclaimed "King of the Number 2." He now looks forward to fishing being the ideal life.

Security Is a Superstition

By William Gurstelle

"Security is mostly a superstition," wrote Helen Keller. "It does not exist in nature ... Life is either a daring adventure, or nothing." In this issue of MAKE, we celebrate daring adventure and working on the edge.

If you read the biographies of some of the best makers of all time, you'll find that a lot of them took astonishing chances when they were young. Nobel Prize winners often reminisce about the time they mixed up a batch of mercury fulminate or ammonium triiodide and found out the hard way that such chemicals can be dangerous. Twelve-year-old Thomas Edison lost his first job when a chemistry experiment went bad and set fire to the baggage car on the train where he worked. David Packard of Hewlett-Packard fame mangled his thumb while experimenting with homemade explosives. When he was only 11, Gordon Moore, founder of Intel Corporation, got caught blowing stuff up with sticks of dynamite.

Of course, times have changed, and 11-year-old boys no longer have access to dynamite, and thank goodness for that! But don't think for a minute that our current era of making things no longer includes danger or adventure. Believe me; working with unfamiliar raw materials, sharp-edged tools, and powerful chemicals presents opportunities for dangerous excitement that rivals hang gliding and scuba diving.

I often hear people lament the idea that we live and work in an overprotective world, a society that discourages people from taking risks and making interesting things. Is it true? Do government officials, insurance companies, and other self-appointed guardians of public safety collude to keep you from making and trying out edgy, fun stuff? I think not.

I've been making and writing about dangerous stuff like propane flamethrowers, 100-pound battling robots with chain saws, and high-powered spud guns for a long time, and I've never had a bit of trouble. In fact, if you don't become a public nuisance or show bad judgment in terms of the time and place of your activities, you are very likely free to try out your ideas in peace.

Truly, there has never been a better time for making interestingly dangerous things. Websites galore sell items ranging from pyrotechnical supplies to torquey traction motors to Tesla coil and rail-gun capacitors. And even better, sources such as MAKE provide detailed directions and advice that make being a 21st-century maker exciting and as safe as possible. (Though Keller was right, there are no guarantees of complete safety.)

So, read on and see what edgy projects in this issue of MAKE call to you. Create a tornado of fire (page 44). Shoot off a dry ice cannon (page 74) or cook up your own rocket fuel (page 68). Finally build that high-voltage Tesla coil (page 46) you've dreamed of. And meet makers like Jon Sarriugarte and Kyrsten Mate (page 38), who are literally playing with fire.

Follow the directions carefully and don appropriate safety equipment. Danger is a state of mind, and with careful preparation, attention to detail, and exercise of common sense, you can be successful and safe every time. ◪

William Gurstelle is a contributing editor of MAKE. A new and expanded edition of his book *Backyard Ballistics* is now available in the Maker Shed.

A maker's life, Nano voltages, MonoBox noise, and tremolo power.

>> In the "Dryer Messenger" project (MAKE Volume 34, makezine.com/go/dryermessenger), I'm confused by the power supply in the parts list. It specifies a 9V power supply, but the Arduino Nano takes 3.3V as input.
—*Charlie Fairchild, Woodstock, Ga.*

AUTHOR THOMAS TAYLOR REPLIES:
The Arduino Nano comes in 3.3V or 5V versions, and many different wall warts will work. Here the Nano's voltage regulator is stepping the 9V supply down to 3.3V. I used the 3.3V Nano because the wi-fi module is a 3.3V-only part. Having them both at the same voltage just made things simple.

>> I like the "MonoBox Powered Speaker" project (Volume 34, makezine.com/go/monobox). The LM386 op-amp is a handy little chip to know about. But the input nags me a little. If you're going to combine the left and right audio outputs of the source device, I recommend some isolation between

them. I would use two 10K resistors, one from L to In+, one from R to In+. Usually a third would go from In+ to Gnd, but the LM386 has this covered by an internal 50K resistor.

Even though the data sheet claims it's optional, a 1µF capacitor in series with the L and R resistors would also help. A switch could go on either side of one cap/resistor for mono or L/R.

I hear what you're saying on R1 being 100Ω for use with a headphone jack, but most of today's devices are meant to drive either headphone or line stages. Not all line-only sources can drive 100Ω without distortion or loss of high frequency.
—*Nick Wallette, Anchorage, Alaska*

AUTHOR ROSS HERSHBERGER REPLIES:
Thanks for the close reading, Nick. We found in testing that some portable headphone players delivered a lot of noise when loaded with high impedance. They need a lot of current drawn to shunt out and damp the high frequency hash in particular. That's why we use the 100Ω input resistor. Terminating the headphone output with a higher resistance will result in higher noise on some source devices.

>> I like the "Optical Tremolo Box" project (Volume 33, makezine.com/go/opticaltremolo) but I wish the power requirement was 9V like most other guitar pedals. It would be nice to plug this into the pedal board's master power supply like all my other pedals.
—*Nick Nicholson*

TECHNICAL EDITOR SEAN MICHAEL RAGAN REPLIES:
You're absolutely right, Nick. I wimped out of including a pulse-width modulation (PWM) motor speed controller in this design and opted instead for a simple series rheostat in-line with the motor. You can only get away with regulating DC motor speed this way when the motor is very low-powered; with a 9V motor the rheostat would get much too hot.

Version 2 of the box will include a proper PWM motor speed control circuit; for instance, there's a simple one based on a 555 timer at dprg.org/tutorials/2005-11a.

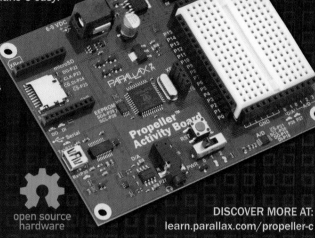

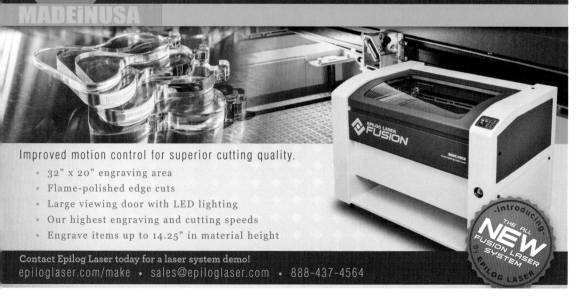

The Moose Notebooks

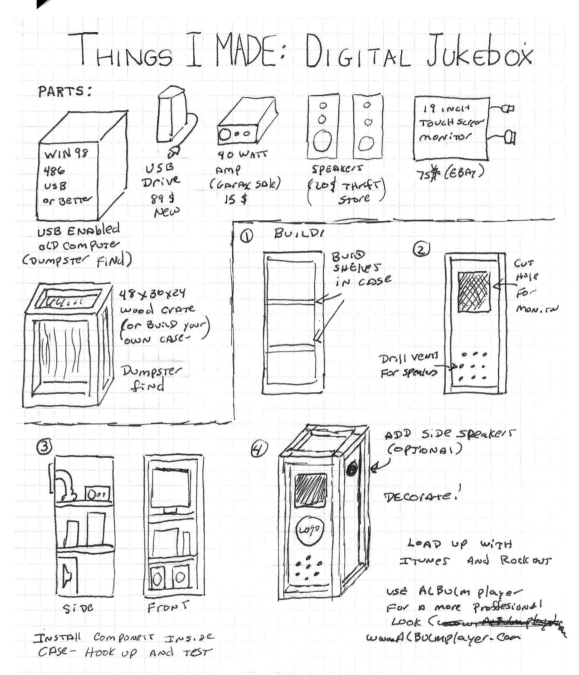

THINGS I MADE: DIGITAL JUKEBOX

PARTS:

WIN 98
486
USB
or BETTER

USB ENabled
oLD CompUTe
(DUMpster Find)

USB
Drive
89 $
New

90 WATT
AMP
(GARAx sale)
15 $

SPEAKErs
(20$ THrift
store)

19 INCH
TOUCH SCreen
moNitor
75$ (EBAy)

48×30×24
wood crate
(or BuiLD your
own case)

Dumpster
find

① BUILD!

BUILD
SHELVES
iN CASE

② CUT
Hole
For
MONITOr

Drill vents
For speakers

ADD SIDE speakers
(OPTIONAL)

DECORATE!

LOAD UP WITH
ITUNES AND ROCK OUT

USE ALBUM player
For A more Proffesional
Look (~~~~~~)
www.ALBUmplayer.Com

③ SIDE   FRONT

④

INSTAll ComPONeNT INSIDE
CASE — HOOK UP AND TEST

SEE THE DEMO! SEARCH yahoo For DIY JUKE BOX moose stUdios

>> I'm writing to thank you for the last seven years. Before MAKE and Maker Faire, I knocked along here in Fresno pretty much alone and largely out of contact with others who have the desire to make oddments. I have been a charter subscriber of MAKE, and my contingent of friends has been attending Maker Faire since the beginning. You know me as the guy who made the Plants vs. Zombies garden pottery sets.

Unfortunately, I cannot attend this year, as I am no longer able to travel. In fact I'm no longer able to work on anything, but I still am able to design things in my head.

I'm sending you a few designs you can pass on to those who might like to build them, including: a digital jukebox made with junk parts; an exercise bike that connects via MaKey MaKey to Google Street View by using its reed switch to produce key presses; and MakerBot-printed parts used to cast bronze statues.

My thanks to Dale Dougherty and company. Your ideas and other efforts have changed my life.

—Sam "Moose" Gueydan,
Clovis, Calif.

EDITOR-IN-CHIEF MARK FRAUENFELDER REPLIES: Sam, thank you for being a part of the MAKE community. We were inspired by your letter. We're sharing your projects here, and with the world at makezine.com/go/moose.

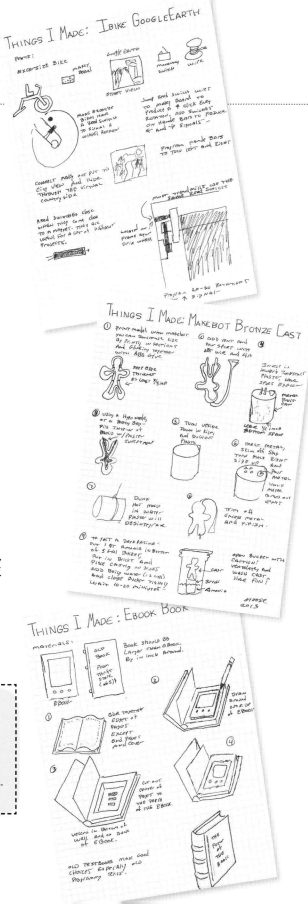

1

The Maker Kids Are Alright

By *Andrew Leonard*

Becca Henry

I want my electric-powered cupcake car and I want it now. A trip to the Maker Faire.

You don't see bicycle-powered cardboard rhinos every day. But when you nearly stumble into one at a Maker Faire, you don't even blink. After just a few hours spent navigating one's way though twirling electric-powered cupcake cars and dashing steampunk-attired ladies and gentlemen, you become well-trained to expect the incongruous and delightfully absurd. And you start thinking, hmm … I've got a lot of cardboard in my basement … what completely bonkers thing can I do with it?

Mark Frauenfelder, editor

in chief of MAKE magazine, calls the Faire "a magic space-time moment." I can't argue. The Maker Faire my son and I visited in San Mateo, Calif., pulsated with a vibe so enthusiastic, so healthy, so creative, and so gosh-darn happy it was impossible not to get swept away. If the hands-on, funky robot, DIY delirious madness of Maker Faire offers us a clue as to where culture is headed this century, then guess what: maybe things aren't going to be so bad after all.

You can learn how to sew or how to make a circuit board at the Maker Faire, play with

Lego or 3D printers, listen to the wisdom of sustainable farming experts, or watch the mad scientists behind Eepy Bird.com drop 640 Mentos into 108 bottles of Coke Zero and create a dazzling ballet of choreographed soda gushers. You can ride a bike seemingly designed by Dr. Seuss, or whirl around a tire swing to the accompaniment of a vintage Pac-Man soundtrack.

You can forge something, should you so please (and if you're wearing your safety-waiver bracelet). Let me repeat: you can forge something. I've taken my kids to

Gunther Kirsch

2

Gregory Hayes

3

Disneyland. You can't forge anything at Disneyland.

Gouts of flame? Check. People on stilts? Check. Motorized couches? Check.

My son and I fought our way into one exhibition space that was so jam-packed and noisy and dark that it was hard to figure out what was going on until we struggled our way deep into the crowd. Then we saw two giant Tesla coils aimed at a cage containing three children. A switch got flipped, bursts of lightning flashed from the coils to the cage, and a seriously rocking version of the theme song to *Doctor Who* started blasting. My son and I turned to each other and high-fived. But the line to get in the cage was too long, so we checked out some drone quadcopters instead.

I've been to geek gatherings — gaming conventions, software programmer conferences — where nearly everybody was white, male, and pale from too much time spent in virtual words. A Maker Faire is geeky as all get out, but (at least in San Mateo) it is also incredibly diverse in terms of age, gender, and race. And everyone shares the same drive: not just to gawk but to get their hands dirty. Everywhere you look, there are stations at which the young and old are encouraged to do something, make something, craft some-

thing. A Maker Faire negates everything you've ever heard about how the ubiquitous spread of computer technology has created a generation of kids locked in thrall to their screens.

Indeed, watching the swarms of kids devouring every possible crafting/hacking/mashup-ing opportunity, I could only think, these are the youngsters who will be opening up their own Etsy stores in the months and years to come. I saw one kid, who couldn't have been more than nine years old, sitting behind a table stacked with old busted circuit boards that had been transformed into analog clocks. "Did you make these yourself?" asked one admiring grandmother. He nodded. If, as William Gibson once wrote, "the street finds its own uses for things," then the Maker Faire is the Rome that all those streets lead to. Repurpose, recycle, reinvent.

In Northern California, the hippie and the geek have long been intermarrying. At Maker Faire, their spawn are coming of age. The Maker Movement is a real thing. And it's wonderful. ◪

4

1. Acme Muffineering's cupcake cars are a beloved staple at Maker Faire Bay Area.
2. Kinetic Creatures' interactive Bike-Powered Cardboard Rhinoceros sculpture.
3. Bright young maker and 3D printing advocate Schuyler St. Leger with his MakerBot.
4. The Crucible industrial arts school presents a blacksmithing demonstration.
5. RadioShack hosted the Learn to Solder tent, where thousands soldered LED badges.
6. The Fun Bike Unicorn Club's pedal car races.

5

6

MAKER FAIRE CALENDAR AUGUST-OCTOBER

WORLD MAKER FAIRE NEW YORK

Sept. 21–22, Queens, N.Y.
Our East Coast flagship Faire is now in its fourth year and growing every year, boasting 55,000 attendees and 650 makers in 2012. Held at the New York Hall of Science, this is the premier showcase of East Coast maker innovation and creativity.
makerfaire.com

MAKER FAIRE ROME

Oct. 3–6, Rome, Italy
Billed as the artisan fair of the 21st century, this first-time event spans four days and will be held at the City of Alternative Economics at Campo Boario in Rome. Collaboratively produced by Rome's Chamber of Commerce with Tecnopolo and Arduino, as part of the World Wide Rome project, it's not to be missed.
makerfairerome.eu/en

JOIN US AT A MAKER FAIRE THIS YEAR AND FIND OUT FOR YOURSELF WHAT THE BUZZ IS ALL ABOUT.

WE GUARANTEE YOU'LL LEAVE INSPIRED.

Andrew Kelly

▼ CERAMICS ▼ GAMES ▼ LASERS ▼ LEGO ▼ PHYSICS ▼ FELTING ▼ TOYS ▼

▼ FIRE ARTS ▼ GADGETS ▼ TEXTILES

▼ **AUG 3**
Maker Faire Hannover (Germany)

▼ **AUG 10**
Rhode Island Mini Maker Faire

▼ **AUG 10-11**
Manchester Mini Maker Faire (U.K.)

▼ **AUG 17**
OC Mini Maker Faire (Calif.)

▼ **AUG 18**
Pittsburgh Mini Maker Faire (Pa.)

▼ **AUG 24**
Dover Mini Maker Faire (N.H.)

▼ **AUG 24-25**
Albuquerque Mini Maker Faire (N.M.)

▼ **SEPT 7-8**
Brighton Mini Maker Faire (U.K.)

▼ **SEPT 14**
Calgary Mini Maker Faire (Canada)

Greenbrae Mini Maker Faire (Calif.)

▼ **SEPT 14-15**
Fort Wayne Regional Maker Faire (Ind.)

▼ **SEPT 15**
Portland Mini Maker Faire (Ore.)

▼ **SEPT 21**
Nashville Mini Maker Faire (Tenn.)

▼ **SEPT 21-22**
Toronto Mini Maker Faire (Canada)

▼ **SEPT 28**
Lewiston-Auburn Mini Maker Faire (Maine)

Tulsa Mini Maker Faire (Okla.)

▼ **SEPT 28-29**
Champlain Mini Maker Faire (Vt.)

▼ **SEPT 29**
Silver Spring Mini Maker Faire (Md.)

▼ **OCT 5**
NoCo Mini Maker Faire (Colo.)

Somerville Mini Maker Faire (Mass.)

▼ **OCT 12**
Charlottesville Mini Maker Faire (Va.)

Emma Willard School Mini Maker Faire (N.Y.)

Groningen Mini Maker Faire (Netherlands)

▼ **OCT 13**
Columbus Mini Maker Faire (Ohio)

▼ **OCT 19**
Cincinnati Mini Maker Faire (Ohio)

Hampton Roads Mini Maker Faire (Va.)

▼ **OCT 20**
East Bay Mini Maker Faire (Calif.)

▼ **OCT 26**
Atlanta Mini Maker Faire (Ga.)

▼ **OCT 27**
Derby Mini Maker Faire (U.K.)

BUSINESS REPLY MAIL
FIRST-CLASS MAIL PERMIT NO. 865 NORTH HOLLYWOOD, CA

POSTAGE WILL BE PAID BY ADDRESSEE

Make:

PO BOX 17046
NORTH HOLLYWOOD CA 91615-9186

Introducing the
Arduino
Robot

By Massimo Banzi and David Cuartielles

After two years of hard work, experimentation, testing, and ups and downs, our robot is ready. The Arduino Robot was presented for the first time during Maker Faire Bay Area 2013. But for us it's not only a matter of launching a new product. It's more important to share a story of how a passion for tinkering is helping us explore new, unexpected roads. That's why I asked David Cuartielles, co-founder of Arduino and the member of our team who has spent more time and sweat on the robot, to reveal how it all happened. In the last three years David moved from knowing nothing about robots to becoming an amateur roboticist with a strong interest in educational robotics. The Arduino Robot is the result of the collective effort from an international team looking at making learning science fun. Arduino is now on wheels. Come and ride with us! —Massimo Banzi

Between 2009 and 2011, I ran an education project at the Computer Clubhouse FARO de Oriente in Mexico City. My role was to bring the craft of electronics to kids 6–18 years old. At the beginning I ran a series of workshops on how to build musical instruments repurposing components we found at the local flea market.

Once the kids became familiar with electronics and programming I asked them what their dream project would be. In the workshop there were both boys and girls, about 25 in total, and independent from their age and gender everybody wanted

to make robots.

To be totally honest, robots are not my specialty, or they weren't back then. I was never really interested in things that moved and performed tasks for me. I thought there were other more interesting fields within electronics than robotics, but I knew I had to follow the kids' dreams at the Clubhouse. It soon became my mission to design an easy-to-replicate robot that could be made using parts existing in Mexico City.

I enlisted Xun Yang, at the time a master's student in interaction design at my laboratory at K3, the School of Arts and Communication at Malmö University, and together we designed a robot that could be easily etched and manufactured by hand. But not only that, we created a whole series of activities for the kids to learn robotics the fun way, from moving the robot to getting it to write text on the floor using a marker. We then made the design open source and Arduino-compatible.

I documented our work on my research blog and got an almost immediate response from the robotics community. It's interesting now to read my diary notes and notice how my opinion about robots was slowly evolving. I started to understand that educational robotics could be a great entry for kids into science. We got a lot of good feedback about what people thought was great in our project and what we could do better. The robotics community is a lively one and people are very willing to share their knowledge.

I researched and bought every book I could find, one of my favorites being *Almost Human: Making Robots Think* by Lee Gutkind. Reading his book I learned a lot about the background of contemporary robotics and came to understand what's most important in the field. One of the stories that caught my interest was the origin of the RoboCup competition as a way to put people's intellect to the challenge of creating the best possible software to solve tasks. The creators went for three different challenges for their world championships:

• **Soccer:** It's a game with equal rules for everybody. It's easy to understand — robots use different techniques to follow a ball and score goals.

• **Rescue:** An activity where the competing teams have to solve a maze-like quest while gathering parts.

• **Dance:** Yes, robots can dance.

One of the many interesting experiences we ran into while working on our project involved a Spanish team called Complubot composed of two kids: Nerea and Iván. Together with their coach, Eduardo, they had been competing — and winning — the Soccer B category at the World Series of the RoboCupJunior (for high-school students).

In a way, RoboCup is like the Formula 1 of robotics. Every year the organization sets a series of rules that make things a little more complex than the previous competition. The teams work throughout the year to make a faster, lighter, and better robot with artificial intelligence. A fascinating rule from the RoboCup is that before entering the competition, each team has to explain its strategy to the other teams to show they've made the hardware and software designs themselves. There is nothing as open source as having to unveil your whole collection of tricks before joining a competition. This is not only about being good technically, but also about being good at explaining how the magic happens. Can you imagine a 12-year-old girl explaining how she, together with her team, built a robot with distributed intelligence by using up to four Arduino Minis and one Arduino Mega? Believe me when I say it's a pretty amazing experience.

When we first met, Nerea and Iván had already won three RoboCup competitions and were on their way to a fourth. We spoke during one of my visits in Madrid, and it became clear to me that we had to work together. They were used to robots using multiple processors that cost as much as $4,000 in parts. Arduino tries to make things as affordable as possible so people can get access to educational tools, therefore my goal became designing a robot that could fit Complubot's needs at Arduino's prices.

I pitched the idea to the rest of the Arduino team, and we started working on the Arduino Robot. The project's code name was Lottie Lemon, named after a character on *The Muppet Show*. I drew the first board that we mounted by hand in early 2011, and we started a long process of iterating designs. Once we had proof of concept, Arduino's hardware guru, Gianluca Martino, took over so I could focus on the software.

During the following year we witnessed a lot of modifications. Every time we solved a bug we found, we came up with a new idea for a feature that could make the robot a little better without compromising the price. The control board went through seven iterations, while the motor board changed nine times. We made seven versions of the operating system until we figured out a way we thought would give people the best introduction to robotics. At the end of the process, we recruited Xun Yang back to the team to create a set of challenges people could undertake to start exploring the world of robotics. The goal was to have fun learning the fundamental operations of the robot, just like the kids who inspired us at the beginning of this story. ◪

Massimo Banzi and David Cuartielles are co-founders of the Arduino project.

✛ Pick up your Arduino Robot today at the Maker Shed (makershed.com).

✛ Check out Massimo Banzi's presentation on Board Building at makezine.com/35.

Barrels of Fun

By *Gregory Hayes* winecountrycraftsman.com

Paso Robles, Calif. artist **Michael Weiss** makes his Winebots out of recycled California wine barrel rings, hand-bent and welded into humanoid shapes and poses. *La Semana*, on exhibit in the tasting room at Wild Coyote Estate Winery, depicts everyman's Monday through Sunday. The installation is only a small sample of the playful characters and scenes Weiss designs, which in turn are a fraction of the art and products Weiss' shop creates, almost exclusively from retired wine production materials, including barrels, vines, and more. Asked if *La Semana* looks like his work week, Weiss laughs, "I work seven days a week. But when you make your living making art, every day feels like the weekend." ◪

These Kids *And Their* Trashy Music

By *Gregory Hayes* facebook.com/landfillharmonicmovie

"A real violin is worth more than a house" in the slums of Cateura, Paraguay, according to local music teacher Favio Chávez. With too few instruments for his students, Chávez found an ally in local luthier and recycler, **Nicolás Gómez**, who began cobbling together instruments from materials scavenged in the landfill surrounding the town. The instruments are surprisingly beautiful, both in appearance and sound. Discovered by documentary filmmakers, the school's Recycled Orchestra has captured international attention. With funding from the Landfill Harmonic Kickstarter campaign, they hope to tour the world and help found similar programs in other impoverished areas. ◪

Courtesy of Landfill Harmonic

Machining the Perfect Controller

By *Sabrina Merlo*

Glenn Black

Glenn Black is a 25-year-old modern-day Renaissance man. He makes quilts; is into parkour, trials unicycling, horizontal tree climbing (tree-to-tree), and skimboarding; is a capable photographer; and plays cello. The San Mateo, Calif. resident is also a hardware/electrical engineer at Sony PlayStation. And in 2010, he joined TechShop to expand his making capabilities.

A self-motivated learner, Black was in the market for a project that would accelerate his experience and mastery of TechShop's wide range of cool tools. One idea was to make his own arcade stick. Black never liked the look and feel of commercially available plastic joysticks. He started playing around in Solid-Works and came up with an inspired shape he describes as, "Kind of a skewed rectangle with a spline along the edge — a very cool curve, something that would be fun to hold."

Black machined the solid walnut control box at TechShop using a ShopBot router. "I had never done a project on the ShopBot before. The overhanging spline was pretty complicated because it required flipping. I had to be really meticulous, think through the steps, plan how to hold my work. And working in wood, you don't have the inherent resolution that you have in metal."

The electronics were comparatively less challenging. Black used a commercially available Cthulhu board designed just for DIY PS3 and PC/Mac joystick making. It comes preloaded with firmware and doesn't even require soldering.

What's next? Black is still exploring what's available to him at TechShop. "In most cases we feel limited by the tools available to us. But now there are no excuses — I'm limited only by my own creativity." ◾

Christian Ristow

High Five

By *Goli Mohammadi* **christianristow.com**

A few years ago, Taos, N.M.-based artist and roboticist **Christian Ristow** was at a festival and felt there was a lack of interactive machinery that prioritized the operator's experience and "violence, of the fun, recreational variety." His answer to these deficits was *Hand of Man*, a 14,000-pound, 26-foot-long hydraulically activated hand and forearm capable of picking up and crushing cars. Participants control the behemoth via a glove-like armature that envelops the user's hand and is designed to accurately trigger *Hand of Man* to reproduce motions made with the glove. Every joint on the hand of the sculpture is a heavy duty steel-on-steel pivot actuated by a solenoid valve. Give a thumbs-up with the glove, and the monster hand gives a thumbs-up back. The sculpture is capable of lifting roughly 3,000 pounds, and while it was designed to pick up cars, it has

also been the demise of pianos, motorcycles, and refrigerators, among other victims.

The initial build took Ristow four months, but the piece has been rebuilt and improved a number of times due to the violent life of a public sculpture. Originally made for Burning Man 2008 and funded with a grant, it has since delighted audiences at two Maker Faires, as well as a number of other festivals. Ristow lists it as a favorite among his portfolio of jaw-dropping sculptures: "Every once in a while you complete a project that just has a certain purity, a clarity of vision, and a conceptual simplicity. And if you can top that off with reasonably good execution, you've got a winner." ◪

To see the power of Hand of Man, check out the video from Maker Faire Bay Area 2013 at makezine.com/35.

David Angnello

The **Light** of **Knowledge**

By *Eric Smillie* eboarch.com/filament-mind

Brian W. Brush

In January, Teton County Library in Jackson, Wyo., flipped the switch on a 70-foot-long sculpture hanging in its lobby. Built from five miles of fiber-optic cable and a tower of 44 LED projectors, the installation uses custom software to link 1,000 separate strands to Dewey Decimal System section numbers. Every time someone queries a subject or reviews a resource on the state's online catalog, the corresponding cable lights up as if a neuron were firing. Called *Filament Mind*, the project is the work of architects **Yong Ju Lee** and **Brian W. Brush**, partners of New York design firm e/b office.

They asked the question, "Can the building play a role as a communicator with the virtual world to exchange information in an active way, in a multidirectional way?" The catalog — which fields roughly 100,000 online queries a month — made an attractive source and a tricky one, because the data it records is basic and anonymous. "I had to find different ways to use the simplified metrics and create something meaningful and interesting," says programmer **Noa Younse**, whose software runs the installation as well as an eye-level display showing real-time searches and historical analysis (most popular category: North American History).

The team planned to hide the LED illuminators, but refraction dimmed the light too much in the long cables, so they were hung in the open. "It was actually a blessing in disguise," says Brush. "It gave us the opportunity to transform these devices into their own kind of aesthetic expression. It's really fascinating to see them at work and exposed." ◼

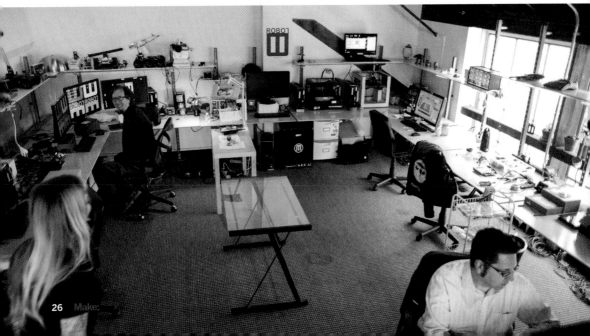

STUPID FUN Club

Written by **Colin Berry**

THESE FOLKS PLAN TO HAVE AS MUCH FUN AS POSSIBLE MAKING TOY ROBOTS, BEFORE THE TOY ROBOTS OVERTHROW CIVILIZATION.

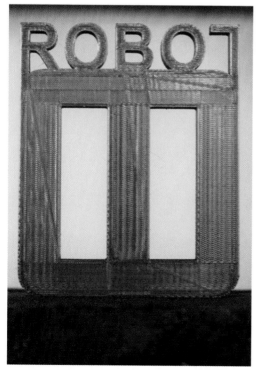

Tucked along an ordinary street in an industrial section of Berkeley, Calif., the company you're about to visit has no receptionist. Instead, at the front door, meet MoonBot, a horny, tractorized robot with a taste for malt liquor; a waiter robot who, years ago, transformed a local restaurant into a testing ground for food-service robotics. (Short version: ketchup bottles got broken.)

Inside and up the stairs, guarded by battle-scarred stars from *Robot Wars*, past a pink box of doughnuts, enter an open-air office where the company's executive team has gathered to load a human-sized trebuchet with water balloons they're hurling at an iPad in order to test a new app that mimics a spectacular smash on the tablet's screen when the balloon "shatters" it.

Ready? *Thhhhhwopp! ... Smash!* Success! Laughter! Applause! It's 10:08 a.m.

Welcome to the Stupid Fun Club (SFC).

A

B

"MAKE NO MISTAKE, WE'RE NOT CREATING CHEAP TOYS HERE. WE'RE CREATING THE SWISS WATCHES OF TOYS."

Founded in 2001 by Will Wright, the game designer behind *The Sims* and *Spore*, and Mike Winter, a Midwestern Gyro Gearloose, robot warrior, and successful startup guy, SFC is a think tank dedicated to exploring the human-machine interface "in a stupid, fun way." For a decade, while tech was nursing its post-boom hangover, SFC worked quietly under the radar, tinkering with (stupid) fun ideas. During that time, as modern DIY and maker culture was born, mobile technologies improved exponentially, providing a platform on which SFC could build the next progeny of toythings. Lately, the thinking SFC has done all along has begun to catapult — er, trebuchet into the public eye.

Besides Mike Winter (Wright now serves as an advisor), SFC's key players include Mike's 26-year-old daughter, Lisa Winter, a brilliant inventor, artist, and former child-celebrity combat roboticist; Mike's sister, Lauren Winter, a savvy systems engineer and user-interface/experience (UI/UX) expert; and Lauren Elliott, a renowned game designer himself (*Where in the World Is Carmen Sandiego?*) and self-proclaimed gadget guy.

SFC is essentially a holding company that spawned a trio of game, robotics, and media ventures, of which only one — Robot 11, which creates devices connected to smartphones and tablets — has emerged from the cloak of secrecy.

The entities cross-pollinate. "We like sharing ideas so the engineers can talk to each other," Mike explains. "We want to make toys that don't require explanation, are gender-neutral, and aren't techy — just turn 'em on and start playing."

SFC's prototypes vary widely: a football that tracks catches and fumbles, as well as how far

A. Arm-mounted vibration motors convey directions to saw-wielding minions.
B. These controls turn a willing human, usually blindfolded, into the RC vehicle for a clamping/sawing machine.

C. Lisa works on a prototype she wasn't allowed to discuss.

and how fast the receiver runs; an animatronic lamp with the AI of a puppy that responds eagerly to a user's movements with soothing sounds and varying brightness; iPhone-based science tools, including a microscope and a crowdsourced bug jar. They also create games, wearables, and robots. Lots of robots. But more about those in a moment.

One favorite gizmo is the Smart Bike, a bicycle that uses GPS, an accelerometer, and sonar to warn its rider if danger is approaching. (A dedicated cyclist, Mike got the idea one day after a car hit Lisa.) A swinging car door? A gaping pothole? The bike communicates the hazard to the rider instantaneously via Google Glass-type goggles. Currently cable-heavy and Bluetooth-enabled, the Smart Bike will eventually be wireless, its controls much more compact.

Gadgets like these, SFC believes, represent the next generation. "Once you get used to being connected, you don't want to go back," Mike says, and he's right: ask anyone who began using a smartphone 10 years ago how much they're dependent on the device. The company sees a direct parallel between the current state of mobile and the historical trajectory of the PC and internet — a shaky start,

followed by a logarithmic boom in capability and performance as tech gets cheaper, better, and ubiquitous. Mobile's explosion and expansion has enabled SFC to create things that they could only have dreamed of just a few years ago.

To produce their prototypes, the company maintains a kick-ass machine shop. SFC shares their HQ with Swerve, a state-of-the-art furniture factory where computerized mill-turn machines transform sheets of laminate and aircraft-grade aluminum into elegant conference tables, swoopy office chairs, and modular shelving. Noisy and chaotic, the plant is reigned over by Ziggy, a bright yellow, 50,000-pound, six-axis robot that dances and whirs with incredible precision, plucking parts from plastic pallets and snapping them into sockets on the mill lathe.

In one corner of the factory floor, Robot 11's machine shop is tiny by comparison, yet well-stocked with high-end lathes and drill presses, belt sanders, sophisticated CNC and other prototyping tools, and a new 3D printer that can transform Tinkercad files and other designs from Thingiverse into real-life plastic objects. Ideas come quickly to life here.

"I love being in the machine shop," confides

Lisa, in goggles, soldering tiny circuits on a stamp-sized Arduino for a wearable she and Lauren, her aunt, are developing. Worn like a watch, the device will eventually be used for geolocation, body monitoring, medical purposes, a smartphone extension, or … who knows?

"We have endless ideas for wearables," Lisa explains. "Some are game-connected, others track your mood, some are just for fun. They need to look good on women, though, like a piece of jewelry that connects to you and makes your life better." She's interested in the "quantified self" movement — tech-based data acquisition systems that track a person's mental and physical performance, inputs, and mood — but says she's disappointed to see so little of it aimed at women.

"I want to make products that help people," she says. "Health, environmental — those are the kinds of things that interest me most. I don't want to just make stuff for entertainment." As it turns out, the company devoted to serious play wants to embrace playful seriousness as well.

Still, this is the Stupid Fun Club, and Lisa's just received a text: "Will's birthday. Come inside for cake."

In the conference room, as someone ties strings to red balloons, Lauren explains how the prototypes she's UI/UXing upstairs will, by definition, be "fun and pretty" and easy for users to learn. The wrist wearable, for example, will probably end up akin to a next-generation pager. She says Bluetooth low energy (BLE) and proliferating freeware have all collided to create the ideal moment in

D. The Waiter Robot, ready to take your order (and break your ketchup bottles).
E. This squirt gun-armed robot uses ping locators to hone in on yard invaders like raccoons.
F. Robot 11's well-stocked machine shop.
G. Lisa finds one of her old *Robot Wars* competition posters.
H. A football that tracks catches and fumbles.
I. The human-sized trebuchet.

toy-making history when — like fashion — everything is coming full circle.

"Except it's a million times more powerful now. And sophisticated," Lauren says. "Make no mistake," she adds, grinning, "We're not creating cheap toys here. We're creating the Swiss watches of toys."

Later, after everyone writes their dreams for the future on the balloons in Sharpie, we step outside to the parking lot to let them go. They rise in a cluster and take a long time to disappear, a glyph of red dots growing smaller and smaller, vanishing incrementally into the pale blue Berkeley sky.

Back in the shop, Mike pulls back a tarp to reveal the heart of SFC's creativity: robots. A Silly String-shooting robot. A robot that draws giant pictures on the ground with chalk. A robot that lights up and wiggles when you dance for it. A flank of tablet-driven combat robots, which will later meet and do battle.

Witnessing this fleet, it's worth recalling Isaac Asimov's 1942 short story "Runaround," which proposes Three Laws of Robotics:

1. A robot may not injure a human being or, through inaction, allow a human being to come to harm.
2. A robot must obey the orders given to it by human beings, except where such orders would conflict with the First Law.
3. A robot must protect its own existence as long as such protection does not conflict with the First or Second Laws.

While the SFC team is evenly divided as to whether or not robots will eventually take over the world (Lisa and Lauren, no; Mike and Lauren Elliott, yes), the bots here conform — for now — to Asimov's rules. Harming one another, however, is another story.

SFC knows a lot about fighting robots: Mike and Will met at *Robot Wars* in the late 90s, as it transitioned from live event to TV show. (It was later canceled, but not before the two founders became known as fearsome warriors with lethal machines.) As a teenager, Lisa was famous for her own fighting robot, Tentoumushi, which (though it appeared a cheerful ladybug) used a "smothering dome" to disable its opponents and propelled her to rank sixth in the world.

It's no surprise then that one of SFC's pet projects is the imminent launch of a crowd-funded line of Lego-based fighting robots. Sold as kits, each bot will contain a cloud-connected ID chip, embedded in a standard Lego figure, that holds the owner's cumulative fighting history. As players get smarter, their bots will, too; plugging the chip into a new bot (or a friend's) will enable it with its operator's accrued abilities. And, like a video game, players will be able to upgrade bots using microtransactions.

Concurrently, SFC plans to reboot a robot combat TV show, supported this time by warriors who can customize their own bots to meet in the ring. Kickstarting the new bot line, SFC believes, will automatically create buzz, and social media will keep the community connected. And once kits get sold, players' nascent ability to print their own plastic replacement parts will keep machines up and fighting.

"The hardest thing about robots is that they're multidisciplinary," Mike says. "For a kid, that can be overwhelming. But now, with the

Image L by Daniel Longmire Photography

Arduino community and the availability of 3D printers, the entire world has changed."

SFC's plan to market kit-based, crowd-funded, cloud-connected, upgradeable combat robots to compete on TV is a perfect recipe for a successful toy. It provides players and audience a healthy serving of competition and aggression — the same reason folks like pro football — yet spices up the enterprise with innovations like cutting-edge tech, DIY subculture, and a pinch of celebrity.

And it can happen fast. "It takes Mattel two and a half years to move a toy from idea to concept," explains Elliott. "We're doing the same thing in days or weeks." The world of play is shifting. "Five years from now," he says, "the top toy companies in the world will be ones you've never heard of today."

● ● ●

It all ends in a brawl. The Winter family — Mike, Lisa, and Lauren — plus Elliott and I are gathered in Elliott's office, where the furniture has been pushed to the walls. Everyone picks a robot. Mine is a thuggish, wedge-shaped bruiser whose lumbering pace, everyone assures me, will be offset by its hefty, hammering cudgel. We plug in our chip-embedded Lego figures, I receive a quick tutorial about my tablet controller, and we're off.

Immediately, Lisa and Mike take control, blithely obliterating Elliott's bot with a few well-aimed blows. Lauren's is next, slaughtered in a whirring smash from her niece's spinning truncheon, and soon the father-daughter pair come for mine. In those last moments before I'm destroyed — before the groans, the laughter, and the splay of Legos splashed like gore across the floor, as I try desperately to deliver a killing strike even as I know I'm done for — it all becomes clear: What Stupid Fun Club is creating every day is nothing short of posterity, nothing less than the rowdy, violent, crude, hilarious, and utterly addictive future of playthings. ◪

+STUPIDFUNCLUB.COM

J. A finned bicycle helmet that lights up when it detects the sound of oncoming traffic.

K. McDonald's made a Happy Meal toy version of Lisa's battle bot.

L. Mike and a teenage Lisa with her award-winning battle bot Tentoumushi.

M. Lego-based fighting robots hold the owner's cumulative fighting history. The chip-embedded Lego robots are controlled via tablet.

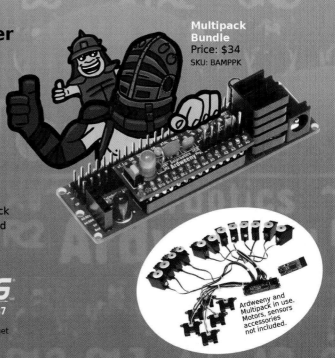

MAKER *Ink*

SEVEN MAKERS WHO ETCHED THEIR DIY PRIDE PERMANENTLY

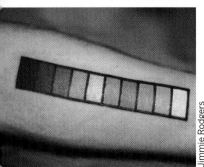
Jimmie Rodgers

Jason Stephens

JIMMIE RODGERS   JANE DAVIS   TYLER WILKINSON

» "I got the first tool tattooed in 1991 and completed the legs in 2012. I tried to get one tool every summer when I could afford it, pulling a tool from my toolbox and taking it to my tattooist, Joe Leonard. Most of the tools were ones that I had 'borrowed' from my father or had been handed down to me from my grandfather. Others were just the old weathered tools that I used every day. All of them are hand tools. Joe is someone I trust, to the point that he would just lay the tool down beside me and start tattooing. After we had filled up both legs with tools, we started on the puppet leg design. My parents were both puppet makers from the 60s to the 80s. When I showed my father the completed legs and pointed out that they were referencing him, he said, 'If you want to honor me, you can just keep your pants on.'"

—Todd Barricklow, Artist
Santa Rosa, Calif.

» "Doing electronics for a living, it's kind of easy to see why I would get such a thing done. However, there's a very small portion of the population that will initially recognize it [resistor color-code chart]. While I didn't really need to have the colors implanted into my skin (already have them memorized), it will be a fun way of teaching others how to read the colors in classes. Also, I had the tattoo made to be 5" long, making each block ½". So I can use this tattoo as a reasonable-guess ruler. ... I created the tattoo image with a Processing sketch, which is open source, obviously."

—Jimmie Rodgers, Hacker + Artist
Somerville, Mass.

» "I got a solenoid valve because I'm a member of the Flaming Lotus Girls [FLG]. I work on the electronics side (I was lead geek for our 2011 project and have been a member since 2009), and solenoids are the intersection of geekery and fire — they're what make the big beautiful poofs of

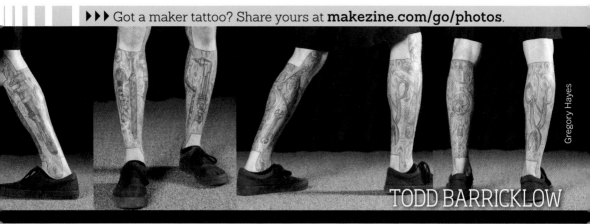

Gregory Hayes

TODD BARRICKLOW

Gunther Kirsch

Greg Flanagan

Sockyung "Sox" Hong

BILL BAYER   GREG FLANAGAN   ZACH SMITH

flame that you see on our projects. My friend Poe has the identical tattoo on his forearm. We were looking for something to commemorate our love of building things, electronics, and working with FLG, and the solenoid was the most fitting thing we could find. Added bonus: If anyone in the shop ever needs a reference diagram, I'm all over that."

—Jane Davis, UX Designer
Oakland, Calif.

» Inspired by his weekend at the Maker Faire Bay Area 2013, Tyler got inked the Monday after. He says, "I was kinda surprised there wasn't a featured tattoo artist at the Faire. That's the only thing I think I didn't see there!" He plans to add a landscaped storyboard around the bot.

—Tyler Wilkinson, Musician + Carpenter
Warwick, R.I.

» "I spent several years as part of a small group of young hackers designing and building radio

electronics under the guidance of two incredibly brilliant senior engineers. We always liked to imagine our project was part of a fiendish scheme for global domination. The missing part of the tattoo is a banner that will someday read, 'With such power I could rule the world.' It's there to remind me of the power of applied knowledge and creativity."

—Bill Bayer, QA Engineer
Oakland, Calif.

» Greg's precisely scaled ruler is exactly 8" from the tip of his pinky to the start of the ruler.

—Greg Flanagan, Fabrication Shop Owner
Minneapolis, Minn.

» MakerBot co-founder Zach has an Ethernet port up his sleeve.

**—Zach "Hoeken" Smith,
Tech Director of Haxlr8r**
Shenzhen, China

BRINGING MAKEY THE ROBOT TO LIFE

Written by **Gareth Branwyn**

Award-winning comic artist Shannon Wheeler joins MAKE.

I first met Shannon Wheeler in the 1990s, at ArmadilloCon, the influential Austin science fiction convention. He gave me a mini-comic of his Too Much Coffee Man (TMCM) strips and I've been a fan ever since. The lovable existentialism of the title character, a layabout not-so-superhero, with a giant coffee cup on his head (who gets his power from caffeine and nicotine) has endured. Shannon still does weekly TMCM strips on his tmcm.com site. He also does work for *The Onion* and *The New Yorker,* and his comics in *The New Yorker* even won him a 2011 Eisner Award.

I was thrilled to hunt down Shannon at the 2012 Comic-Con, where we enthusiastically started discussing the possibility of him doing a regular comic for MAKE. He was a fan of the magazine already and was anxious to explore the idea of working with us.

The result is Shannon's take on Makey the Robot, the iconic Maker Faire mascot designed by Kim Dow. Shannon explains his idea behind the character:

> *"All of us are a combination of competence and fallibility, enthusiasm and self-doubt."*

"All of us are a combination of competence and fallibility, enthusiasm and self-doubt. Makey is a robot eagerly trying out new projects. I want to use him to express the conflicting feelings of creation we experience by using a fallible robot struggling with his own creations. I want to keep the dialog to a minimum to emphasize the emotional dimensions of our well-meaning hero. I want the cartoons to be poetry more than parody, a quiet commentary on the creative process. In future strips, Makey will go on many building adventures, deal with fellow robots, and meet other makers. Hopefully he won't break or blow up too many things along the way."

Look for additional episodes of Shannon's "Makey" as the end page comic in future issues of MAKE. And perhaps we can even coax the ham-clawed one (Makey, not Shannon) to appear on the website from time to time, too. ◪

Gareth Branwyn writes on art, DIY media, and technology, and is the former editorial director for Maker Media.

MAKEY

SHANNON WHEELER

DREAM IT.
MAKE IT!

END

Playing with Fire

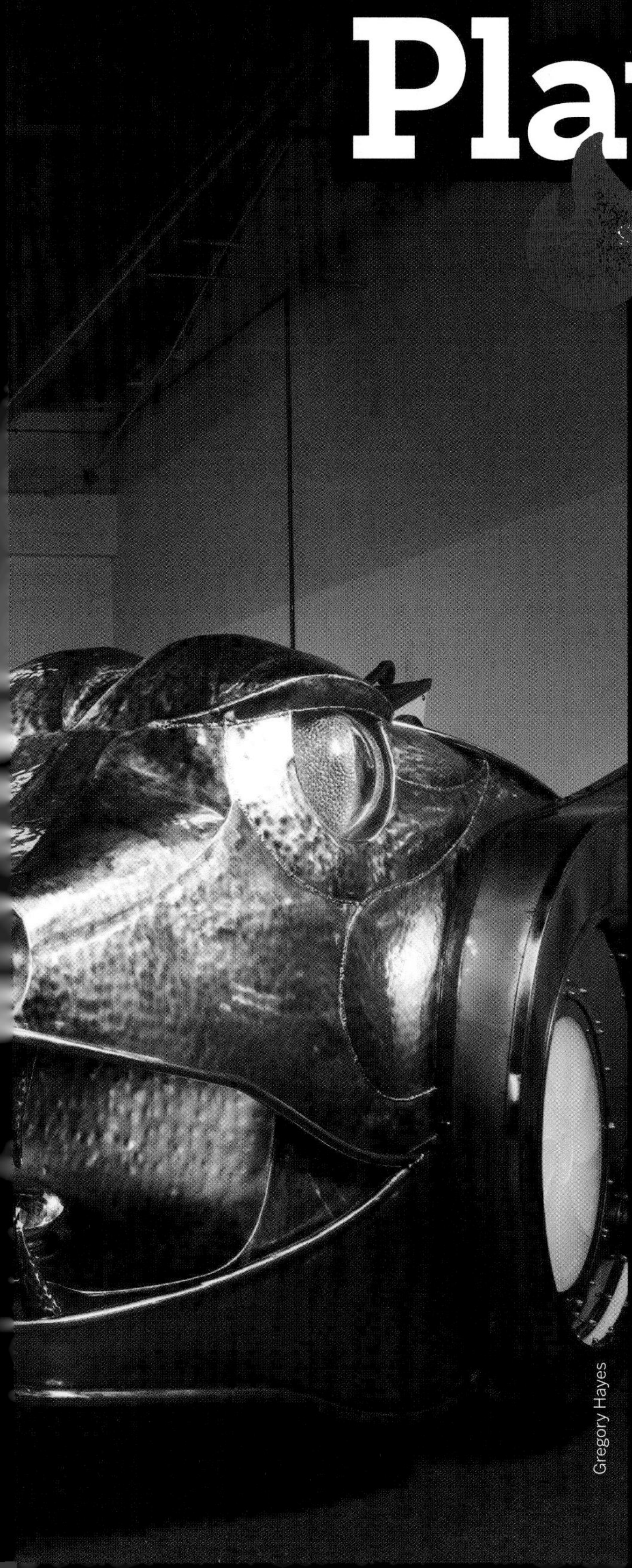

Meet makers who are not afraid to turn up the heat.

Written by Goli Mohammadi

Jon Sarriugarte and Kyrsten Mate are a Bay Area maker couple whose creativity knows no bounds, as evidenced by their stunning *Serpent Twins* 50-foot kinetic mobile sculptures on this issue's cover. They've exhibited projects at Maker Faire Bay Area since 2006 along with their daughter Zolie, now 8, who has grown up with the Faire. Jon recently gave us insight into the inspiration and tech specs for the *Twins*.

Gregory Hayes

Gregory Hayes

What was the inspiration behind the Serpent Twins? At Burning Man (BM) 2010, we met some people from Northern Europe who wanted us to build a car for their festival. They have lots of water at the festival and I started thinking about a serpent powered by a jet ski. I remembered the factory down the street in West Oakland was throwing out green 55-gallon drums, which would work great for the body.

Kyrsten and I discussed it on our way home from BM, but by then we knew the Europe trip was underfunded, and our discussions turned to BM ground-based cars. I searched for months for the right base vehicle and finally came across two electric Taylor-Dunns. We had bounced back and forth from LED blinky body to steel and fire, and came up with the idea to do both! Yin and yang, his and hers, old school and new. Glad we did both.

Tell us your design process from concept to creation. We started with concept drawings done by Tansy Brooks and myself. Kyrsten found and compiled historical pictures and images, and over the span of several months, distilled the look we wanted. For the head and body, no shop drawings were made. We worked off the concept sketches, made life-sized cardboard models, and mocked up the head in pipe cleaners! Once the basic shapes were done, Kyrsten made paper patterns for the sheet metal, which were then transferred to steel and shaped by hammer and stump to fit our frame.

The body barrels were the most challenging, as every one was unique. What we thought would be an easy production line turned into 20 one-off trailers. The tail was sketched out on a napkin for our crewmates Toast [Jeffrey McGrew] and Jillian [Northrup] from Because We Can. Toast modeled it in CAD, I had my local metal supplier cut it on their plasma cutter, and in less than a week Toast and Jillian had the tails done just in time for Burning Man.

How long did they take to build? We call our crew the Empire of Dirt. This team, headed up by Kyrsten and myself, has many wonderful and talented people with skills in electrical engineering, welding, gaming, film, software, design, carpentry, arts, smithing, auto, and more. Without these people, their input, time, expertise, funds, and love for building, these creatures would not be here. I can't think of a better family to work and play with!

I want people to really understand how much goes into a project like this. It took a little over four months over the summer with a crew of 40 people to build the *Serpent Twins*. We had a team making the head, barrel, and fins, converting and wiring the tow motor/head, and a team building the LED lighting and controls. All of these teams came together in the last weeks as the various pieces came out of the spray booth and final assembly took place. We have gone back now every summer to upgrade the lighting and repair various problems.

What's under the hood? We used a Taylor-Dunn as the base vehicle. It's a 24-volt electric tow motor originally from the NUMMI auto plant. We have 400 amp-hours of batteries with a backup generator in the tail and a 50-amp spine of power that runs the length of the beast. Propane is stored in the second barrel in the black serpent and feeds the head for the fireball generator that shoots a ball of flame 20+ feet. A 400-amp car stereo and 18" subwoofer with dual three-way speakers make the beast roar to life when one of Kyrsten's custom serpent sounds is played.

The 15,000 RGB LEDs are controlled by a BeagleBoard in the head that talks to the custom "Serpent Charmer" PSoC boards designed by Keith Johnson though a WIZnet card. Each barrel acts as an individual network device pulling its code out as the packets pass by, refreshing and sending the rest to the remaining barrels. The math-based patterns are controlled from the cockpit though a MIDI controller with drum pads and knobs. We are able to choose from a variety of patterns and manipulate their speed, color, and pop flashes throughout the body.

This year the code team added a tablet interface, and we are currently designing a UI to teach kids and adult how math-based graphics work. This UI will allow the user to play with the numbers and see the changes they make right on the serpent. If you're interested in helping, the code can be found at serpenttwins.com.

Being a husband and wife creative team, what is your collaboration process like? It's challenging to keep our crew together, give our daughter the attention she wants, keep a house, work full time, and find spare time to do projects. Kyrsten works for Skywalker Sound cutting and mixing movie sound and I have my own company, Form & Reform (formandreform.com), where I design lighting for home and restaurants, and I now have several dozen custom lights at Restoration Hardware. As hard as it is to make the time, once we're in full swing the creative juice flows, and as you can see we make some wonderful things together.

Does your daughter Zolie help with the builds? She has, and she continues to be more and more a full-fledged team member. She brings love and understanding to the meetings, introduced hand raising and talking in turns, and comes up with simple, fun ideas. She does some metalwork, including pounding out the fins on the *Twins*. You will find her in the shop with us most weekends working with someone and being very involved with our projects. She enjoys the crowds and performing as well.

■ ■ ■

Fire Makers

SMOKIN' HOT: [Clockwise from above] *Gon KiRin*, the 60' long, 26' tall dragon that spews over 10' of flames, made by Ryan C. Doyle and Teddy Lo. Photographed at Burning Man 2012, *Gon KiRin* has been to three Maker Faires. Lucy Hosking and her *Satan's Calliope* fire organ at the inaugural Maker Faire Bay Area in 2006. *Chester the Fire-Breathing Horse* art car, built by Jason Anderholm and Rebecca Anders, at Maker Faire Bay Area 2013. The Flaming Lotus Girls' *Cochlea* methanol fountain at Maker Faire Bay Area 2012.

ise from top: Cliff E

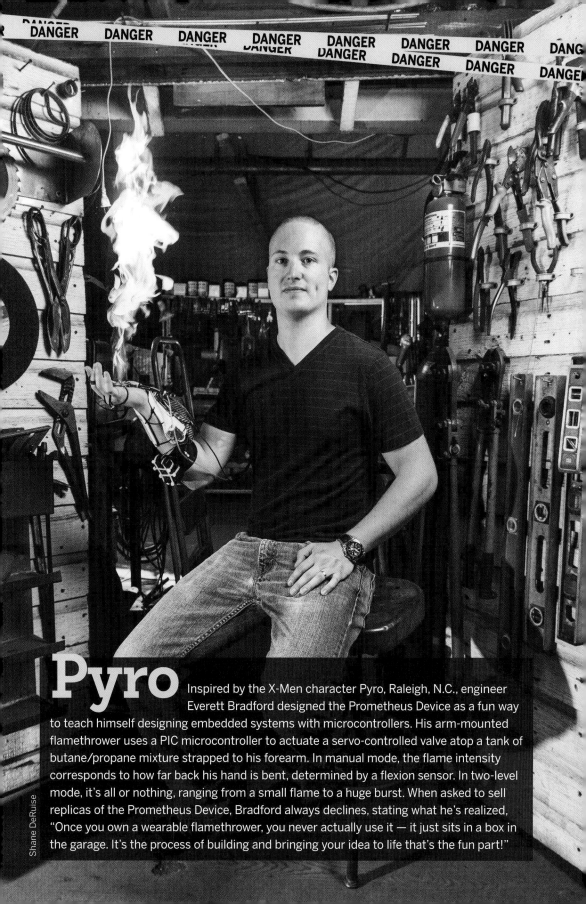

DANGER DANGER DANGER DANGER DANGER DANGER DANGER DANG
DANGER DANGER DANGER DANGER DANGER
DANGER DANGER DANGER DANGER DANGER

Pyro
Inspired by the X-Men character Pyro, Raleigh, N.C., engineer Everett Bradford designed the Prometheus Device as a fun way to teach himself designing embedded systems with microcontrollers. His arm-mounted flamethrower uses a PIC microcontroller to actuate a servo-controlled valve atop a tank of butane/propane mixture strapped to his forearm. In manual mode, the flame intensity corresponds to how far back his hand is bent, determined by a flexion sensor. In two-level mode, it's all or nothing, ranging from a small flame to a huge burst. When asked to sell replicas of the Prometheus Device, Bradford always declines, stating what he's realized, "Once you own a wearable flamethrower, you never actually use it — it just sits in a box in the garage. It's the process of building and bringing your idea to life that's the fun part!"

Shane DeRuise

DANGER DANGER DANGER DANGER

⁄ **TIME: 1–2 HOURS**   ⁄ **COST: $20–$40**

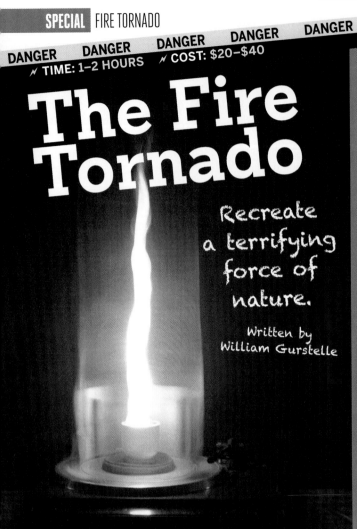

The Fire Tornado

Recreate a terrifying force of nature.

Written by William Gurstelle

MATERIALS

» **Modeling clay, ½lb**
» **Phonograph turntable** Check thrift stores, garage sales, and repair shops. Look for an old one with a 78rpm setting and a raised edge on the platter (see Step 2). If you can't find one that does 78rpm, it's OK: you can easily modify a belt-drive turntable to increase its speed.
» **Crucible, or fireproof bowl or cup, teacup-sized**
» **Crucible or fireproof bowl, large** to fit completely over the small one
» **Epoxy, fast drying; or hot glue gun and glue**
» **Craft sticks, 6"×¾" (12) or tongue depressors**
» **Window screen, aluminum, 36"×36"** not plastic or fiberglass
» **Straight pins (4–8) or narrow gauge wire**
» **Kerosene, 2tsp** *Do not* substitute any other fuel such as gasoline or alcohol.
» **Cotton rag, approximately 4"×4"**

TOOLS

» **Tinsnips or aviation shears**
» **Lighter, long handled**
» **Fire extinguisher, general purpose**

⚠ This project is to be performed by responsible adults or under their closest supervision. Keep careful watch on the fire at all times. Remove all combustible and flammable objects from the area. Keep the project away from flammable walls, surfaces, curtains, etc. Do not allow the craft sticks or anything else to ignite. This project creates smoke and fire and should be done outdoors. Keep your fire extinguisher close by.

Fire whirls often happen during wildland fires. They're usually small, but under the right conditions, big ones form. Huge whirls have been recorded climbing to heights over 3,000 feet, with wind speeds exceeding 155mph — equivalent to an F3 tornado.

Meteorologists call any such phenomenon a *vortex*, which is just a volume of rotating fluid. Examples include smoke rings, water going down a drain, and dust devils.

In fact, fire whirls and dust devils are very similar. Both form when a layer of cool air passes over a layer of lighter hot air, which pokes a hole in the cool layer and rises through the opening. If conditions are right,

the rising air begins to spin and a whirlwind forms. The rotational trigger can be as simple as a gust of wind.

The difference between dust devils and fire whirls is mostly a matter of degrees — temperature, that is. Where the dry earth under a dust devil can be as hot as 150°F, air temperatures in a fire whirl can exceed 2,000°F.

These extreme temperatures create huge

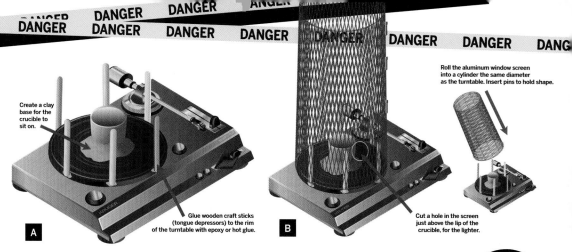

Create a clay base for the crucible to sit on.

Glue wooden craft sticks (tongue depressors) to the rim of the turntable with epoxy or hot glue.

A

Roll the aluminum window screen into a cylinder the same diameter as the turntable. Insert pins to hold shape.

Cut a hole in the screen just above the lip of the crucible, for the lighter.

B

columns of rising hot gas that shoot up erratically through cooler layers. Some small triggering event sets the column spinning, and a fire whirl is born. The hot gases spin like a hurricane, with greater wind speeds closer to the center, or eye. The suction pulls in extra oxygen, intensifying the flames.

Build the Fire Tornado

The Fire Tornado dramatically illustrates the effect air has on a fire's shape, burn rate, and fuel consumption. It's easy to build.

1. Use modeling clay to create a base for the crucible in the center of the turntable, so it stays put while spinning.

2. Epoxy or hot-glue the craft sticks to the lip of the turntable (**Figure A**) and let dry. (I glued mine to a raised edge midway across the platter, with a diameter of about 8½".)

3. Roll the aluminum window screen into a 36"-high cylinder of the same diameter, and cut as needed to overlap the edges by 1". Secure the seam with straight pins or wires through the mesh.

4. Carefully place the cylinder over the sticks so they hold it vertically in place. Cut a small access flap in the screen right above the lip of the crucible (**Figure B**). If the screen wobbles too much when it spins, try securing it to the sticks with pairs of super magnets.

Use the Fire Tornado

Working outside, with a fire extinguisher within reach:

1. Drip 2tsp kerosene onto the cotton rag.

2. Put the rag on the turntable.

3. Secure the screen over the craft sticks.

4. Put the lighter through the flap and

ignite the rag. Note the flame's size and shape.

5. Start the turntable at its highest speed. What happens to the size and shape of the flame?

Put Out the Fire

1. Stop the turntable.

2. Put on your gloves and remove the screen.

3. Invert the large bowl and cover the crucible, smothering the flame.

NOTE: The vortex is best viewed at night and in calm wind.

TIP: If your turntable's top speed is 33rpm or 45rpm, remove the belt's drive pulley, shave it down a little smaller, and reinstall. My modified turntable clocks at least 65rpm. (The MAKE Labs team replaced their pulley with a smaller 3D-printed one.)

Fire Tornado Physics

The flame produces hot gases that rise from the center of the burning materials, displacing cooler air above, which then sinks down around the edges of the cylinder.

Because the cylinder is spinning, the centrifugal motion pushes this cooler air toward the rotating screen, making the relative air pressure at the screen higher. But if the pressure is higher at the screen, it must be lower somewhere else, namely in the hot area above the flame. The lower pressure there allows hot gases from the burning rag to rise more easily, creating the flame vortex of the fire tornado. ↗

William Gurstelle is a contributing editor at MAKE. This article is excerpted from his book The Practical Pyromaniac (Chicago Review Press).

William Gurstelle (facing), James Burke (A and B)

↗ **TIME: 40–60 HOURS** ↗ **COST: $100–$200**

The Six-Pack
Tesla Coil

Written and photographed
by Craig Newswanger

Build a
classic
spark-gap
coil that
can throw
15" arcs
without
pricey
commercial
caps.

Nikola Tesla's inventions are all around us: radio, AC power, fluorescent lighting, and remote control are just a few. He was ahead of his time, and his work with high-frequency alternating currents has inspired engineers, scientists, inventors, artists, and (frankly) quacks for more than a century. The Tesla coil is particularly fascinating because of the elemental, visceral nature of the electrical arcs it produces. It's like watching lightning strike. Tesla used these spectacular effects to wow audiences with the wonders of AC electricity.

MATERIALS

» **Neon sign transformer, non GFI, 9kV/25mA** Any transformer outputting 9kV–12kV at 25mA or 30mA is suitable, but if it's not 9kV/25mA, you may have to alter the coil design using the JavaTC app (see "Designing the Six-Pack Coil").

FOR THE SECONDARY COIL:
» **PVC pipe, 1½", 24" length**
» **PVC pipe caps, 1½", flat type (2)** Lasco #447015RMC
» **Magnet wire, 32 AWG, ½lb** All Electronics #MW-32-2
» **Scrap of shim brass or copper, 0.005"–0.01" thick**
» **Screw, brass, ¼-20×1"**
» **Hex nut, brass, ¼-20**
» **Solder lug, ¼"**
» **CPVC pipe cap, ½"**
» **Brush-on gloss urethane finish** I used Minwax Fast-Drying Polyurethane.
» **Plywood or MDF, ½" or ¾", 16" circle**
» **P-clip, small**

FOR THE SECONDARY COIL WINDING JIG:
» **Lumber, 2×4, 24" length**
» **Corner brackets, 3" (4)** Stanley #DPB115
» **Threaded rods, ¼-20×24" (2)** aka all-thread
» **Drywall screws, fine thread, #6×1" (8)**
» **Hex nuts, ¼-20 (4)**
» **Fender washers, ¼"×1¼" (3)**
» **Split lock washers, ¼" (2)**

FOR THE TOP-LOAD:
» **Wreath form, foam toroid, 12"×1¾"**
» **Plywood or medium-density fiberboard (MDF), ¼"×10"×10"**
» **T-nut, ¼-20×⅜"**
» **Aluminum tape, 2"** such as 3M #3381

FOR THE SIX-PACK CAPACITOR:
» **Glass bottles, 12oz, long neck (6)** I used Bud Ice bottles.
» **Solid copper wire, 12 AWG insulated, 6'**
» **Stranded copper grounding wire, 18/1 bare, 6'** or strip a lamp cord
» **Table salt, 2 cups**
» **Mineral oil, about 10fl oz**
» **Coil spring, tension, small**
» **Stranded copper wire, 12 AWG, insulated, 48"**
» **Vinyl drip hose, ¼", 30" length**
» **Alligator clip**
» **Crimp-on lugs, ring-tongue (2)**

FOR THE PRIMARY COIL:
» **Plywood or hardboard, ¼"×24"×24"**
» **MDF, ½"×6⅛"×13"**
» **Wood screws, brass, flat-head, #4×1" (16)**
» **Solid copper wire, 12 AWG insulated, 60'** It's cheaper if you buy a 100' spool.
» **Solid copper wire, 6 AWG bare, 50"** or use

» ¼" copper refrigeration tubing
» **Cable ties, small (8)**
» **Masonite or birch plywood, ⅛"×4"×5"**
» **Vinyl drip hose, ¼", about 4"**

FOR THE PRIMARY COIL SLOTTING JIG:
» **Lumber, 1×4, about 18"**
» **Steel pin, ⅛"×1½"** e.g. a roofing nail with its head cut off

FOR THE SPARK GAP:
» **Acrylic or polycarbonate sheet, ¼"×3"×8"**
» **Copper pipe couplers, ½" (7)** Nibco #600
» **Machine screws, brass, #6-32×⅜", round head (5)**
» **Hex nuts, brass, #6-32 (10)**
» **Split lock washers, #6 (5)**

FOR THE TERRY FILTER:
» **Acrylic or polycarbonate sheet, ¼"×8"×10"**
» **Corner braces, brass, 1" (8)**
» **Screws, brass, ¼-20×1¼", round head (4)**
» **Hex nuts, brass, ¼-20 (8)**
» **Cap nuts, brass, ¼-20 (4)**
» **Machine screws, zinc plated or brass, #6-32×1" (12)**
» **Machine screws, brass, #6-32×½" round head (4)**
» **Split lock washers, #6 (12)**
» **Nuts, zinc plated or brass, #6-32 (20)**
» **Crimp-on lugs, ring-tongue, #6×16 AWG (18)**
» **Stranded copper wire, 16 AWG insulated, about 3'**
» **Resistors, wire-wound, 1K, 100W (2)** Vishay #71-HL100-06Z-1.0K
» **Varistors, metal oxide, 1,800V, 2,500A (14)** Panasonic #P7215-ND
» **Capacitors, film, 3,300pF, 1.6kV DC (14)** Panasonic #P10501-ND
» **Resistors, carbon film, 10MΩ, ½W, 5% (14)** Yageo #10MH-ND
» **Vinyl drip hose, ¼", 5½" length**
» **Lumber, 1×2, 20" length**
» **Wood screws, brass, #6×1" flat-head (4)**

FOR THE MOMENTARY SWITCH:
» **Extension cord, 3-wire, 25'**
» **Household electrical box, 1-gang, with grounding screw and blank cover**
» **Romex clamps (2)**
» **Crimp-on lugs, ring-tongue (2)**
» **Crimp-on butt-splice connector, insulated**
» **Switch, SPST, pushbutton, 2A 120V AC**

FOR FINAL ASSEMBLY AND GROUNDING:
» **Stranded copper wire, insulated 10 AWG or heavier, 10' or more** to reach a ground electrode or grounded water pipe
» **Wire nut, large**
» **Stranded copper wire, insulated: 12 AWG or heavier, several feet; and 18 AWG or heavier, several feet**
» **Crimp-on lugs, spade- or ring-tongue (10)**
» **Vinyl drip hose or Tygon hose**

TOOLS

» Miter box and saw
» Sanding block, small
» Sandpaper: 220 grit, 600 grit
» Steel wool
» Combination square
» Pencil
» Center punch
» Cordless electric drill An extra battery pack will come in handy.
» Cable ties for holding drill trigger
» Twist drill bits: $\frac{1}{16}$", $\frac{7}{64}$", $\frac{1}{8}$", $\frac{1}{4}$", $\frac{5}{16}$", $\frac{3}{8}$"
» Brad-point drill bits: $\frac{1}{16}$", $\frac{1}{8}$", $\frac{9}{64}$", $\frac{1}{4}$"
» Countersink bit
» Phillips driver bit
» Hole saw, $2\frac{3}{8}$" (optional)
» Rubbing alcohol
» Contact adhesive, Goop brand
» Cyanoacrylate (CA) glue aka super glue
» Wood glue
» Paintbrush, $1\frac{1}{2}$"
» Electrical tape
» Marker, permanent felt-tip aka Sharpie
» Ruler
» Cotton glove (optional)
» Soldering iron and solder
» Epoxy, J-B Weld
» Scissors
» Utility knife
» Compass or trammel points capable of 8" radius
» Wrench, $\frac{1}{4}$-20 bolt, and flat washer for seating the top-load T-nut
» Jigsaw
» Measuring cup
» Liquid funnel
» Resealable plastic jug, 72oz
» Multimeter with capacitance setting
» Wire cutter/stripper/crimper
» Pliers, small
» Computer, printer, and paper
» Table saw with miter gauge
» Table saw blade with $\frac{1}{8}$" kerf
» C-clamps (2)
» Kitchen oven
» Round wooden dowel, 1" or larger for smoothing coil wire
» Propane torch or heavy-duty soldering iron
» Painter's blue masking tape
» Tinsnips
» Vise
» Camera tripod or similar adjustable stand

Since Tesla's time, hobbyist "coilers" have made many discoveries and improvements to the basic design, achieving bigger sparks with less input current. With the advent of plastics, improved wire insulators, and a better understanding of theory, the modern Tesla coil looks very different from the original. The basic circuit and concepts are the same, but almost everything else is different.

One thing that is the same, in this project, is the capacitor design. Ours is made from glass beverage bottles, very similar to the champagne bottles that Tesla himself often used.

Safety

Along with the wonder and awe of a Tesla coil comes a significant level of danger. It is the responsibility of anyone who builds or operates a Tesla coil to ensure the safety of themselves and anyone who might come near, either during a demonstration or inadvertently. Whenever you approach the coil, unplug the power cord and hang on to the plug end as you work. If the location is not entirely secure, consider adding a safety key switch so you can pocket the key.

A Tesla coil's high-frequency electrical field can damage or destroy cardiac pacemakers/defibrillators, hearing aids, and other biomedical devices. I've never seen this happen, but it's imperative to warn audiences of the possibility before demonstrations.

Similarly, the Tesla coil can damage other sensitive electronics nearby. I have personally destroyed a stereo receiver, a garage door opener, a wireless phone system, and two PC network cards. It again falls to the maker to make sure that the coil is operated at a sufficient distance from any valuable electronics, flammable materials, pets, and of course, small children.

There are many hazards to be aware of, and in this single article we cannot cover them all. If in doubt, contact a nearby Tesla hobbyist or an engineer experienced in high-voltage devices and electrical safety. *If you have any doubt about your abilities in this area, don't attempt the build.* Period!

How It Works

Fundamentally, a Tesla coil is just a transformer, like the one that steps household electricity down to a voltage suitable for charging your cellphone. All transformers have two coils — a primary and a secondary — and most of those you encounter in daily life transform voltages based on the different numbers of turns in each coil. A Tesla coil works on a slightly different principle, creating the very high voltages needed to produce long arcs in open air mostly through the *inductive difference* between its primary and secondary coils.

DANGER!

↗ **Assume the capacitor is always charged.** Capacitors can retain a charge for days. No matter what anyone else tells you, always safely discharge the six-pack capacitor yourself, and jumper it with a sturdy clip lead before touching *any* of the components. Keep the jumper in place when you're not operating the coil.

↗ **Do not operate the coil around small children or animals.**

↗ **Operate in clear spaces at least 20 feet from flammable materials.** The electric field generated by a Tesla coil can create sparks within furniture and in the ceilings of structures. Sparks can ignite combustible solids, liquids, and especially vapors.

↗ **Do not touch the NST terminals.** Both sides of the neon sign transformer are "hot." Some NSTs have exposed primary terminals carrying line voltages. Current at the NST secondary terminals is

usually low, but the voltages are high enough to cause painful shocks and secondary injuries from loss of motor coordination.

↗ **Do not stare at the sparks.** Electrical arcs in air emit ultraviolet light that can damage eyes and skin on extended exposure. Clear polycarbonate sheet can be used to shield the spark gap and block most of the UV generated by the sparks.

↗ **Do not operate the coil without proper ventilation.** Electrical arcs in air produce ozone, nitrogen pentoxide, and several other nitrogen oxides that are hazardous to health. Note that nitrous oxide is *not* produced.

↗ **Do not operate indoors without ear protection.** This Tesla coil can produce hazardous levels of noise. It's less of a problem outside, but indoors the sound is *loud*.

Six-Pack Tesla Coil Schematic

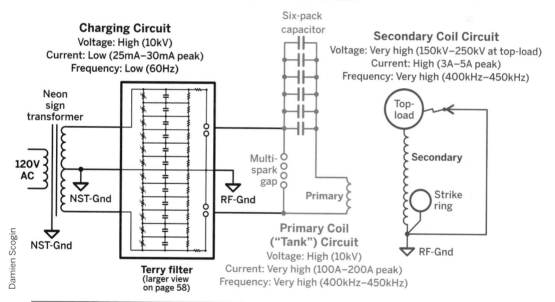

More specifically, a Tesla coil is an *air-core, dual-resonant* transformer. *Air-core* means the coils are hollow, not wrapped around metal or ferrite cores as in common transformers. *Dual-resonant* means the circuits containing both primary and secondary coils are tuned to "ring" at the same frequency.

The combination of the primary coil (an inductor) and the capacitor (the bottles, in this design) create a resonant LC circuit that "rings" at a particular frequency. This is called the *tank circuit*. Since both tank circuit and secondary coil are tuned to the same frequency, they pass

energy back and forth when "struck" with an electric impulse. Imagine striking a bell near a drumhead tuned to the same note.

The electrode on the top of the coil is called the *top-load*. You can imagine it as a capacitor with one side connected to the secondary coil, the other side connected to ground, and the air all around as the insulator between the two "plates."

This Tesla coil is powered by plugging into a wall outlet, and uses a neon sign transformer (NST) to step 120V AC up to about 10kV at 25mA–30mA. Solid-state voltage

Six-Pack Tesla Coil Anatomy

Strike ring
Attracts arcs that might otherwise hit primary coil.

Six-pack capacitor
Charged by NST until spark gap fires.

Top-load
Adds capacitance and allows voltage to increase even further without "leaking" as corona discharge.

Secondary coil
Has distributed capacitance due to adjacent winding. Rising/falling magnetic field from primary coil induces very high voltage here.

Neon sign transformer (NST)
Boosts line voltage to roughly 10,000 volts, but at a lower current.

NST protection filter ("Terry filter")
Recommended to keep high frequencies from damaging NST.

Spark gap
Voltage-sensitive switch, closes when air breaks down and arcs.

Primary coil
Receives large current impulse when gap fires, creating large magnetic field.

RF ground
(*Not* outlet ground.)

James Burke

converters are not appropriate, nor are modern NSTs manufactured with ground fault protection circuitry. You'll need a used or old-stock NST; fortunately these are not hard to find on eBay and, sometimes, Craigslist. Neon shops may have old units hanging around.

Designing the Six-Pack Coil

The math for designing a Tesla coil is not especially difficult, but it can get tedious if you want to iterate to get the best design. Fortunately, coil hobbyist Bart Anderson has paved the way with a wonderful JavaScript program called JavaTC. If you're interested in the math, Bart's site classictesla.com has resources and links that will lead you as deep as you want to go.

JavaTC was instrumental in designing the six-pack Tesla coil. The output text file describing the six-pack coil is available from makezine.com/35.

Spend some time playing with JavaTC, tweaking the specs for the six-pack coil, and you'll see how the various design parameters affect one another. If you have to use a different transformer, make a different top-load, use a different wire gauge or any other major changes, you can use JavaTC's auto-tuning feature to understand how to modify the design.

Build Your Six-Pack Tesla Coil

First-time "coilers" should follow this build as closely as possible. Use a neon sign transformer rated for 9kV at 25mA, strive for a main tank capacitance as close to 0.005µF as possible, and do not substitute parts if it can be avoided.

Plan your build carefully before you start. Don't just jump in and start building without reviewing every aspect of the design. High-frequency resonant circuits are very sensitive to small changes, and poor attention to planning can make tuning the coil a frustrating job.

Craftsmanship is also important. Take your time, particularly with the secondary coil, where a single crossed winding or a skimpy varnish job can easily result in a nonfunctional or short-lived coil.

Good design, attention to detail, and patient craftsmanship will pay off with a long, noisy spark that draws oohs, aahs, applause, and admiration from everyone who sees it.

1. Wind the Secondary Coil

1a. Cut a 16" length of 1½" PVC pipe. Clean up the ends by sanding.

1b. Mark and drill a ¼" hole in the exact center of both PVC caps.

1c. Build the coil form and a simple winding jig

1b

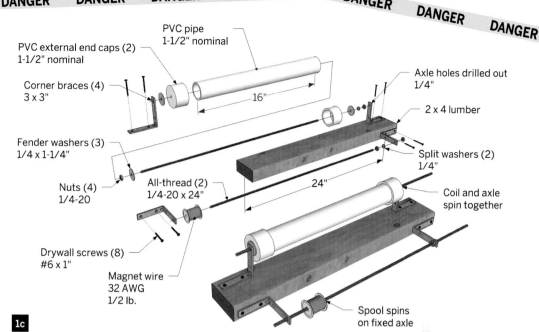

PVC pipe
1-1/2" nominal

PVC external end caps (2)
1-1/2" nominal

Corner braces (4)
3 x 3"

Axle holes drilled out
1/4"

16"

2 x 4 lumber

Fender washers (3)
1/4 x 1-1/4"

Split washers (2)
1/4"

Nuts (4)
1/4-20

All-thread (2)
1/4-20 x 24"

24"

Coil and axle
spin together

Drywall screws (8)
#6 x 1"

Magnet wire
32 AWG
1/2 lb.

Spool spins
on fixed axle

1c

as shown. Clamp or otherwise mount it right at the edge of your work surface.

1d. Chuck the secondary coil's axle into your drill.

1e. Clean the surface of the PVC pipe with rubbing alcohol and dry it carefully.

1f. Make sure your drill has a full charge before winding the coil. If you have a spare battery, keep it handy. Using your finger (or a cable tie) to hold the trigger, spin the tube slowly.

1g. Apply an even coating of brush-on gloss urethane to the tube. To avoid drips and runs, leave the tube spinning until the urethane dries a bit. This first layer of varnish helps hold the wire in place during winding. After 20 minutes the urethane should still be a bit tacky.

1h. Stop the drill. Use electrical tape to fasten the end of the magnet wire to the cap at one end of the tube.

1i. Use a Sharpie to mark the winding start- and end-points at 2" and 14½" from the starting end of the tube, including the cap.

1j. Set the direction of rotation so that the top of the tube rolls away from the wire spool. Use your nondominant hand (or a helper) to run the drill slowly as you guide the wire onto the tube with your dominant hand.

1k. Advance the wire with wide turns until you reach the first witness mark, then start feeding so that the wire is smoothly *close wound*, with

1d

1g

NOTE: You'll remove the PVC end caps later, so be careful not to varnish the cap/pipe seam.

1i

NOTE: To guard against friction burns, wear a cotton glove or use a folded paper towel to guide the wire through your fingers.

1j

1k

Do not allow the wire to cross a previous turn. If it does, stop, carefully unwind the coil, and correct the mistake. A few small gaps between turns will not matter, but a single crossover will create a short circuit that renders the coil useless.

2a

NOTE: Place the brass strip carefully. CA glue bonds quickly. ⟶

2d

2e

1l

NOTE: No matter what it says on the varnish can, do not sand between coats since this will likely damage the wire insulation and ruin the coil. Also, when handling the finished coil be careful not to ding or scratch the windings.

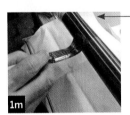

1m

2f

2g

no gaps. The trick is to angle the incoming wire so that it's slightly *behind* the advancing edge of the coil. Be patient and careful. Keep winding until you reach the second witness mark.

1l. Close-wind the wire a few turns past the second witness mark, then open up the winding again until you reach the second end cap. Wind the wire about 5 turns on the end cap and tape it securely.

TIP:
The whole winding process should take less than 20 minutes. The varnish will hold the winding in place if you need to rest for a minute, but don't let it dry completely or it won't be sticky enough for you to finish.

1m. Take a short break and change the battery on your drill, if necessary. As before, set the drill spinning slowly with a cable tie across the trigger. Apply at least 3 coats of gloss urethane over the secondary windings, letting each coat dry according to the recoat time listed on the can. To prevent drips and runs, keep the coil spinning while each coat is first drying.

1n. Set the coil aside to dry thoroughly. Don't even think about using it until the last coat of varnish has dried for 24 hours.

2. Install the Secondary Coil Terminals

The coil terminal connections must be mechanically sound both at the bottom ground wire and at the top-load.

2a. To make the ground terminal, remove the tape from one end of the secondary coil and unwind the wire right up to the first witness mark. The wire should release easily from the varnish. Remove the pipe cap.

2b. Cut a ½"×2" strip of shim brass or copper. Polish both sides with fine sandpaper or steel wool until they're bright.

2c. Form the strip so that its curve matches the PVC pipe, using your fingers or a small-diameter tube or dowel.

2d. Apply a 4" piece of electrical tape to the outside/convex side of the strip and a sparse amount of CA super glue to the inside/concave side. Carefully glue the strip in place about ⅛" below the winding, as shown, using the electrical tape to hold it as the glue sets.

2e. After a few minutes, remove the electrical tape and tin the edge of the strip nearest the coil, where you'll connect the wire. Cut the wire so that it wraps smoothly around the pipe to contact the strip without kinks or bends. Clean the end of the wire using 600-grit sandpaper, then tin it and solder it to the strip. Finally, apply a dab of Goop adhesive to immobilize the wire near the connection.

2f. To make the top-load terminal, unwind the other end of the coil to the witness mark and remove the remaining PVC pipe cap. Pass a ¼-20×1" brass screw through the hole from the inside. Place a ring-tongue solder lug over the screw, apply a brass nut, and tighten well.

2g. To insulate the screw head and prevent arcing down the inside of the tube, epoxy a ½" CPVC pipe cap inside the 1½" cap, as shown.

2h. Use PVC cement or epoxy to glue the PVC end cap back in place on top of the coil. Once the glue has set, wrap the loose end of the coil wire around the pipe and cap in a tidy, gentle spiral approaching the solder lug on top. Avoid sharp bends. Tape the wire in place temporarily, and cover it with Goop from the top witness mark to the top edge of the cap.

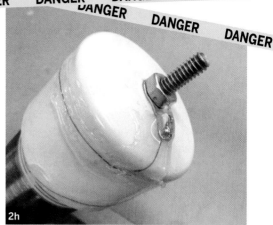

2h

After the glue sets, remove the tape, clean the end of the wire with 600-grit sandpaper, tin it, and solder it to the lug. Cover the final length of wire, from the edge of the pipe cap to the solder lug, with more Goop.

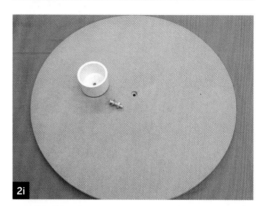

2i

2i. Cut a 16" circle of ½" or ¾" MDF or plywood. Drill a ¼" hole in the center, and counter-bore on one side so that the head of a ¼-20×1¼" brass machine screw sits flush.

2j. Mount the loose PVC pipe cap to the base using the machine screw and a matching brass nut inside the cap.

2k. Push the secondary coil into the PVC cap. Solder a 6' ground wire (18–12 AWG) to the metal strip on the bottom end of the coil. Secure the wire to the base with a P-clip for strain relief.

3. Build the Top-load

The top-load is based on a toroid-shaped floral wreath form I bought at a large hobby store. If you can't find a foam toroid just like this one, there are other options for building top-loads, but these variations will affect your tuning. Use JavaTC to understand how a different top-load may affect your design.

2j

CAUTION: Do not drill through the secondary coil form (the PVC) to attach the ground wire. Doing so may cause internal arcing.

2k

3a. If your toroid has mold lines or other protrusions, smooth them all out with a sanding block.

3b. Measure the inside diameter of your toroid. Mine was 8¼". Use a large compass or trammel points to draw a matching circle on a sheet of ¼" or ½" plywood or MDF. Cut the circle slightly oversize with a jigsaw, then sand it to fit snugly. Drill a 5⁄16" hole in the center of the disk.

3c. Cover both sides of the disk with strips of aluminum tape, overlapping them by about ¼". Burnish it with the side of your Sharpie to smooth out wrinkles, then trim it flush with the disk's edges and center hole using a utility knife.

3d. To make sure the disk is centered vertically inside the toroid, first set the toroid on your work surface, then arrange small blocks of wood inside to support the disk at the correct gluing height. With the specified toroid, ¾"-thick blocks work nicely.

3e. Apply super glue to the inside edge of the toroid, then quickly press the disk into place.

3f. Wait a few minutes, then run a bead of super glue along the joint for good measure.

3g. When the super glue sets, completely cover the toroid in 8" strips of aluminum tape. Each strip should wrap all the way around the toroid and extend onto the top and bottom of the disk.

3h. Work each strip of tape down smoothly with your fingers to eliminate big wrinkles and voids. Then burnish it smooth before applying the next piece of tape. Overlap the strips about ¼" on the outside of the ring.

3i. Install a ¼-20 T-nut in the hole with a dab of epoxy for good measure. Seat the nut firmly by tightening it against a matching bolt and washer. Then remove the bolt and washer and set the top-load aside while the epoxy sets.

4. Build the Six-Pack Capacitor

A capacitor is a device for storing energy, in the form of electric charge, between 2 conducting electrodes separated by an insulator, aka a *dielectric*. In this case, salt water inside the bottles is one electrode, the bottle glass is the dielectric, and the outer foil covering is the other electrode. There are a number of homebrew ca-

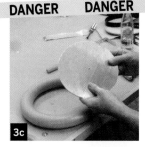

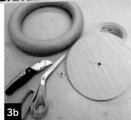

pacitor designs, but the bottle capacitor, aka *Leyden jar*, is by far the simplest and easiest to make. (*For another project using Leyden jars, see MAKE Volume 17, "The Wimshurst Influence Machine."*)

4a. Wash the bottles thoroughly with soap and water. Dry them.

4b. Apply 2 pieces of aluminum tape, crossed at 90°, to the bottom of each bottle. Use scissors to trim the tape in a circle about ½" larger than the bottle's base, then snip inward to create small triangular "petals." One by one, fold and smooth the petals up along the bottle's sides, working carefully to reduce wrinkles. When all the "petals" are in place, burnish the tape smooth.

4c. Cover the sides of each bottle with overlapping strips of aluminum tape, each of which wraps all the way around the bottle's circumference, plus 1" or so. Start flush with the bottom edge, covering over the "petals" from the previous step, and work your way up, overlapping the strips by ¼" to ½" and burnishing each strip before applying the next.

TIP: Fold each strip of tape in half lengthwise, sticky sides out, to find its center. Then align the fold with the vertical center of the toroid's outer edge.

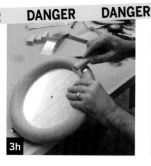

3h

3i

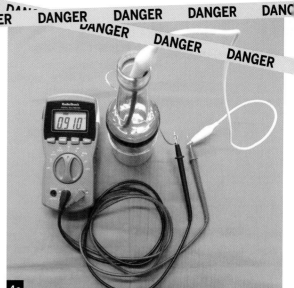

4b

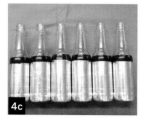

4c

4e

4f

4g

4h

4k

Apply the last strip so that it ends just where the bottle curves inward at the neck. Wrap the top edge of the foil with 2 turns of electrical tape to reduce corona discharge.

4d. Thoroughly mix 2c table salt into 72oz warm water in a resealable container.

4e. Use a funnel to fill each bottle with saltwater to just below the top edge of the foil tape, then use your multimeter to measure its capacitance as shown here. This coil is designed for a primary tank capacitance of 0.005µF. The bottles will be connected in parallel, so in an ideal 6-bottle capacitor, each bottle will have a capacitance of 0.005µF / 6 = 0.00083µF, or 0.83nF.

Add, remove, or swap out individual bottles as needed to bring the total capacitance as close to 0.005µF as is practical. You can also adjust the capacitance of an individual bottle, by peeling a bit of foil tape from the top edge or by removing some saltwater.

When you're done testing, empty the saltwater solution back into your resealable container.

4f. Tape the bottles together tightly in groups of 3, using 3 or 4 turns of vinyl electrical tape at

both top and bottom, as shown.

4g. Tape the 2 sets of bottles together as shown.

4h. Wind a length of bare stranded wire twice around the group of bottles, crimp ring-tongue lugs on each end, and stretch a small coil spring between them to keep the wire tight against the foil.

4i. Cut an 18" lead from 12 AWG insulated stranded copper wire, strip the ends, and solder one of them to the bare winding. For strain relief, secure the lead to the winding a short distance away using a small cable tie. Cover the remaining length of the lead with vinyl drip hose, for extra insulation, then crimp a ring-tongue lug to the loose end. This lead will connect to the spark gap.

4j. Fill each bottle with saltwater as before, then top with a ⅛" layer of mineral oil to prevent corona discharge.

4k. Connect the saltwater electrodes inside the bottles using a wire dip electrode made from two 19" lengths of 12 AWG solid copper wire, each wrapped around a third 23" length, as shown. Bend the 6 leads down, where neces-

5a

5b

sary, to reach the bottoms of the bottles. Make good mechanical connections and solder the joints well.

4l. Cut a 30" lead of insulated stranded 12 AWG wire, strip the ends, and solder one of them to the dip electrode. Run the lead through a length of ¼" vinyl drip hose to add extra insulation, then solder an alligator clip to the loose end. The alligator clip will be used to adjust the primary coil tap point during the tuning stage.

5. Build the Primary Coil

JavaTC is useful for calculating the number of turns in the primary coil given a particular NST, secondary coil configuration, top-load geometry, and main tank capacitance. If you use an NST rated at 9kV/25mA, build the secondary coil and top-load as described above, and achieve a main tank capacitance of 0.005μF (± 0.0002μF), this primary should work nicely. If any of those factors differs significantly in your build, use JavaTC to check whether the primary coil design needs to be adjusted, before proceeding.

5a. Use a large compass or trammel points to mark a 16" circle on ¼" plywood or Masonite. Preserve the center mark; you'll need it later. Cut out the circle with a jigsaw.

5b. Cut 8 strips of ½" MDF measuring 1½"×6⅛". These will be the combs that hold the flat spiral of wire that forms the primary coil.

5c. You'll cut the wire slots on a table saw using a simple spacing jig. Build the jig, as illustrated, from a piece of 1×4 lumber and a suitable ⅛" steel pin, super- or hot-glued in place. I used a decapitated roofing nail.

5d. To cut the first slot, clamp the jig onto your saw's miter gauge or panel sled. Mount a saw blade of about ⅛" kerf and set it to cut ⅜" deep. Set the MDF comb blank against the jig, covering the slot, with its end against the pin. Activate the saw and cut the first slot in the comb.

5e. To cut subsequent slots, simply advance the comb along the jig, index the pin in the preceding slot, and make the cut. Proceed in this manner, cutting 15 slots in each comb.

5f. Cut 8 strips of ⅛" Masonite or birch plywood 3½"×½" to make the strike ring supports.

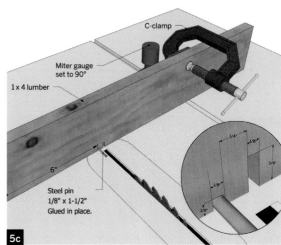

C-clamp

Miter gauge set to 90°

1 x 4 lumber

6"

Steel pin 1/8" x 1-1/2" Glued in place.

1/4"   1/8"

3/8"

1/8"

5c

5d

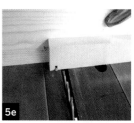

5e

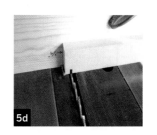

5f

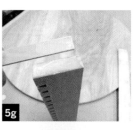

5g

5i

TIP: If you're in a humid area, dry the combs in an oven for 1/2 hour at 250°F before varnishing. Apply the urethane while they're still warm.

5j

5k

5l

![warning] The comb screws should be solid brass, not plated. In large, high-power coils, steel screws can get hot enough to burn loose from the primary coil form. This is called inductive heating.

5n

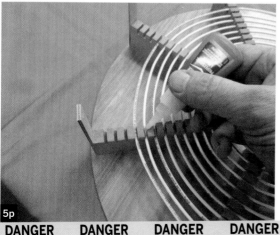

5o

5p

NOTE: I arranged my combs in a very shallow spiral, with each comb 1/16" farther out from center than the one before. This isn't necessary for the coil to function, but it makes the coil spacing a little neater.

Drill a ⅛" hole near the end of each, to pass a zip tie.

5g. Affix a strike ring support to one end of each comb with wood glue.

5h. Give the combs a coat of urethane varnish.

5i. Use a pencil and ruler to divide your disk into 8 equal "pie slices." The angles don't have to be exact; eyeball-accurate is fine. Mark 2 screw holes on each line, at 3¾" and 6" out from the center.

5j. Center-punch and drill ⁷⁄₆₄" holes at your 16 marks, and countersink them from one side so that a #4 wood screw sits flush.

5k. Cut a 2⅜"-diameter hole in the center of the disk using a hole saw or jigsaw.

5L. After they've thoroughly dried, mount each comb on the disk with two #4 brass wood screws. To prevent splitting, drill a ¹⁄₁₆" pilot hole for each screw before tightening it into the MDF. Your coil form is complete.

5m. To wind the primary coil, you need about 60' of 12 AWG insulated solid copper wire. First smooth out any kinks by fastening one end to a secure object (like a trailer hitch or signpost), wrapping the wire once around a wooden dowel at least 1" in diameter, and pulling through the whole length, letting the wire slip around the dowel as you go.

5n. Drill a ¼" hole in the plywood disk near the first comb slot (at the center of the spiral).

5o. Pass about 12" of wire through to the underside of the coil form, then start laying wire into the slots in the combs, forming a smooth spiral. Continue until there are 4 turns remaining unwound.

5p. Go back and seat the wire you've already laid firmly at the bottom of each slot, then apply a drop of super glue to keep it in place.

5q. Wind the last 4 turns temporarily and mark them between each pair of combs as shown. You'll strip the insulation from short sections of the wire, and these should be staggered to prevent accidental shorts between adjacent turns.

5r. Remove the outer 4 turns of wire from the form and use wire strippers to cut the insulation where marked. Put the wire back in the slots, and use a utility knife to slit and remove the small sections of insulation between the cuts.

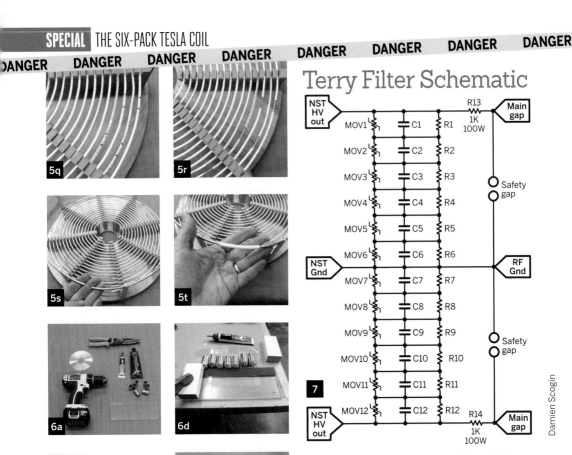

Terry Filter Schematic

Damien Scogin

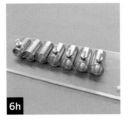

also use copper refrigeration tubing. Make sure to leave a small gap between the ends of the ring, or it will act like a shorted turn in the primary and ruin the coil's performance.

5t. Slip a short section of vinyl drip hose over the free ends of the strike ring, positioning it to cover the gap. Use a propane torch or heavy-duty soldering iron to solder a 6' length of insulated stranded wire, 18 AWG or heavier, to a point on the ring near a support. Run the wire down the support and away from the coil for later connection to RF ground.

6. Build the Spark Gap

"Coilers" have produced many designs for spark gaps over the decades. This is a simple, static "multi-gap" spark built from ½" copper pipe couplings and an acrylic or polycarbonate baseplate. It is designed to be adjustable between 1 and 6 gaps, depending on where you attach the leads. With the specified 9,000-volt NST, you will probably use 4 or 5 gaps.

JavaTC tells us we want a gap of about 0.042" between our couplings. CD-ROM plastic is about 0.043" thick, so pieces cut from an old

Seat and secure with super glue as before.

5s. Arcing between the top-load and the primary coil is potentially dangerous, and bad for the coil. Our design includes a strike ring mounted above the primary to shunt these wayward arcs directly to ground, like a lightning rod. Cut a single loop of 6 AWG bare solid copper wire and attach it to the strike ring supports with small zip ties through the holes. You can

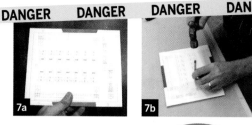

TIP:
Use brad-point bits to avoid cracking the plastic, and drill into a sacrificial piece of MDF or plywood to keep the bit from binding as it breaks through.

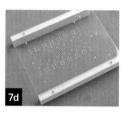

disc will make handy spacers.

6a. Cut a 3"×8" plate from ¼" acrylic or poly-carbonate sheet. Drill holes in the corners for later mounting to a wooden base. Use a brad-point bit to avoid cracking the plastic.

6b. Remove any sticky labels from the pipe couplings. Make holes in 5 couplings by clamp-ing each one in a vise, marking a dimple with a center punch about ⅜" from one end and drill-ing a 3⁄16" hole through one side only.

6c. Use tinsnips to cut 6 spacers roughly 1"×1" from an old CD-R.

6d. Run a piece of blue painter's masking tape along one long edge of the plate, as shown, to serve as a guide when gluing the pipe couplings. Clamp a block of scrap wood to one end of the plate, with its inside edge 1⅜" from the end.

6e. Apply a bead of epoxy to one of the drilled couplings, opposite the hole, and affix it to the plate, against the wood block, with the hole facing up.

6f. Insert a CD-R spacer and glue one of the undrilled couplings to the plate beside the first. Be careful not to get glue on the spacer. Repeat, using the second undrilled coupling.

6g. Epoxy the 4 remaining drilled couplings to the plate, holes pointing up, with spacers be-tween. Clamp a second wood block against the

last coupling to hold it all while the epoxy sets.

6h. Remove the blocks and spacers. Install a #6-32×⅜" brass machine screw in each drilled coupling from the inside, add a split washer and a nut outside and tighten firmly. You'll add an additional nut when you connect a lug.

7. Build the Terry Filter

Tesla coils are hard on neon sign transformers. I have killed several over the years. Coil hobbyist Terry Fritz (capturedlightning.org) understood that the high frequencies and high voltages of a Tesla coil could cause arcing inside an NST, leading to the formation of "carbon tracks" which eventually ruin it. In the late 1990s, he published designs for filter network circuits to go between the NST and the rest of the Tesla coil circuit to prevent this harmful feedback.

The resistors and capacitors form an RC filter that attenuates high frequencies from the Tesla coil. The resistors bleed high voltage from the filter caps, and will also bleed the charge from the main tank capacitor if the NST should fail. The chained metal oxide varistors (MOVs) form a "surge suppressor" that shunts big voltage spikes — for instance from the top-load arcing to the primary coil — straight to ground.

7a. Cut an 8"×10" plate from ¼" acrylic sheet. Download the layout template from makezine.com/35, print it at 100% size and tape it to the plate at the edges.

7b. Use a center punch to transfer the hole locations from the printout to the acrylic.

7c. Drill the holes where indicated, using a 1⁄16" bit for the caps, resistors, and varistors, and a 9⁄64" bit for the screws.

7d. Cut two 10" rails of 1×2 lumber and attach the plate to them using four #6×1" brass flat-head wood screws.

7e. Attach a 1" corner brace at each end of the 2 large wire-wound resistors as shown, using a #6-32×½" brass machine screw, 2 brass #6-32 hex nuts, a #6 split washer, and a #6 flat washer to attach each brace.

7f. Assemble 4 safety gap terminals. Each terminal consists of a corner brace, a ¼-20× 1¼" brass round-head machine screw, a ¼-20 brass cap nut, and two ¼-20 brass hex nuts.

7f

7g

7i

7j

7k

7l

7m

NOTE:
When you wire the screw terminals, add a #6 split washer between the lug rings and the hex nuts to keep the threads from vibrating loose.

TIP:
The Terry filter has 6 terminals for connecting other parts of the Tesla coil circuit: 2 NST power leads, the NST ground lead, the RF ground lead, and the 2 sides of the main spark gap. Assemble these terminals with the hex nuts on top of the baseplate as shown, not underneath.

9a

9b

7g. Assemble the 2 sides of the RC filter circuit from 14 pairs of resistors and capacitors. First, insert the leads of a capacitor where indicated by the template, through the holes in the baseplate. Then insert a 10MΩ resistor's leads, bending them as needed, beside the capacitor. Now bend each resistor lead over, wrap it once around the adjacent capacitor lead and clip off the excess, connecting the 2 components in parallel. Proceed to the next resistor-capacitor pair, until you've finished a row of 7 pairs. Repeat for the other side of the RC filter.

7h. Now bend one lead of each capacitor over, wrap it once around the near lead of the adjacent capacitor, and trim the excess, connecting all 7 RC pairs in series. Repeat for the other side of the RC filter.

7i. Insert a chain of 7 varistors, where indicated, beside the 7 RC pairs. Bend one lead of each varistor over, wrap it once around the near lead of the adjacent varistor and trim the excess, connecting all 7 varistors in series. Bend the free leads of the first and last varistors over, wrap them once around the free leads of the first and last capacitors in the adjacent filter circuit and trim the excess, connecting one varistor chain and one side of the RC filter circuit in parallel. Repeat for the other side of the circuit.

7j. Solder all connections securely. Leave the first and last capacitor leads on each side of the circuit intact, but cut away any remaining excess leads.

7k. Mount the 2 large, wire-wound resistors and the 4 safety gap terminals to the baseplate, as shown, using #6-32×1" brass machine screws and matching hex nuts.

7L. Referring to the schematic, photos, and assembly template, complete the Terry filter circuit using short jumpers made of 16 AWG insulated stranded wire and terminated with matching #6 ring-tongue lugs. Reinforce the insulation on the long NST ground jumper by running it through a length of vinyl drip hose.

7m. To connect the RC filter circuits to the wire-wound resistors, simply crimp lugs to the capacitor legs and bend them over to the resistor screw terminals. The filter-ground connections will require short jumpers soldered to the capacitor legs.

7n. After assembling the Terry filter, set each

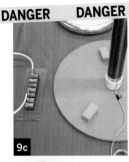

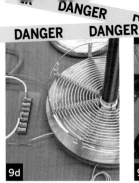

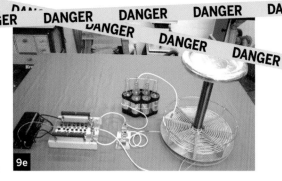

9c

9d

9e

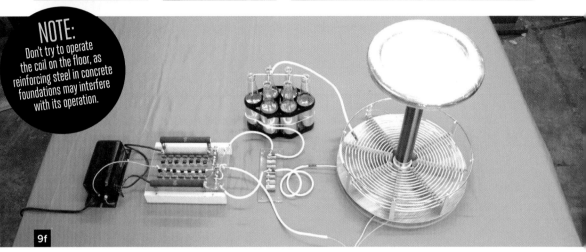

NOTE:
Don't try to operate the coil on the floor, as reinforcing steel in concrete foundations may interfere with its operation.

9f

safety gap to ⅛" by adjusting the bolts and nuts, using a ⅛" drill bit as a gauge. You'll adjust this distance later so that the gaps only fire if a high-voltage spike occurs. See makezine.com/35 for the detailed procedure.

8. Build a Momentary Power Switch

If you don't want to be plugging and unplugging a cord to turn the coil on and off, you'll want to build a simple momentary switch. See makezine.com/35 for instructions.

9. Assemble the Coil

9a. Set up on a non-metallic table, *without metal components*, in an area safe for operating the coil. (See "Danger" warnings, page 51.)

9b. Terminate the NST's factory leads with #6 ring-tongue lugs to fit the Terry filter's terminal screws. Connect the NST "hot" outputs to the large wire-wound resistors. Connect the NST output ground to the Terry filter grounding terminal. Don't forget to add lock washers between the nuts and the lug rings.

9c. Super-glue 3 small spacer blocks cut from ¾" wood or MDF to the base of the secondary

coil, as shown.

9d. Slip the primary coil down over the secondary coil and set it on the spacer blocks. Be careful not to scratch the secondary windings. Run the primary coil lead out through the space between the primary and secondary coil baseplates.

9e. Align the brass screw on top of the secondary coil with the T-nut in the top-load, and rotate the top-load to thread them together. Do not overtighten.

9f. Referring to the photos and schematic, connect the remaining coil components using stranded leads. The wire used in the charging circuit and the secondary circuit can be narrower (no lighter than 18 AWG), while the wire in the tank circuit must be thicker (no lighter than 12 AWG) to handle the large impulse currents there.

Use the shortest possible leads to minimize losses, and reinforce wire insulation wherever possible by running leads through vinyl drip tubing or, with heavier gauges, clear Tygon hose. Avoid contact even between insulated wires. Make good connections and terminate all leads with crimp-on spade (fork) or ring-tongue lugs.

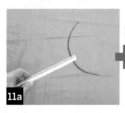

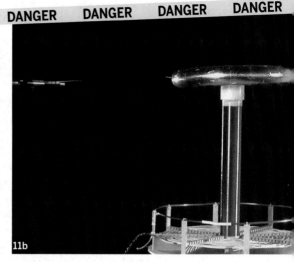

10. Ground the Coil Safely

10a. Run a length of insulated 10 AWG (or heavier) stranded copper wire from the Tesla coil to a metal water pipe that you know is buried in the ground or to a dedicated ground electrode near your electrical service, and connect it securely.

10b. As indicated in the schematic, gather the ground leads from the secondary coil, the strike ring, and the Terry filter. Add to these 3 the ground lead from your top-load strike point (see Step 11b), and connect all 4 together using a large wire nut, with the RF ground lead running to your water pipe or other ground electrode.

> ⚠️ Do not operate a Tesla coil without a proper ground. Do not use the ground connection on an electrical outlet. Instead, connect an independent "RF ground."

11. Tune the Coil

Once the coil is assembled and properly wired in a safe place, you can begin to tune. Theoretically speaking, the goal is to bring the resonance of the tank circuit to the same frequency as that of the secondary coil circuit; practically speaking, the goal is to find the point on the primary coil to attach the alligator clip from the six-pack cap that gives the longest spark from the top-load. A test oscillator and oscilloscope can be used to tune the coil, but if you don't have those tools, this empirical, methodical approach should yield good results:

11a. Make a capacitor discharge tool so you can safely short out the six-pack capacitor between experiments. It's just a 10" length of

12 AWG or heavier solid copper wire mounted on a handle of PVC pipe. To discharge the cap, touch the wire to the dip electrode and the outer foil simultaneously — keeping your fingers clear!

11b. Tape a short bit of bare small-gauge wire to your top-load to act as a breakout point. Set up a tripod or other device to hold a grounded wire near the breakout point. This is your strike point.

11c. Hook up the spark gap so that only 2 gaps are in use.

11d. Connect the alligator clip lead from the six-pack capacitor to the outermost point on the primary coil.

11e. Set the ground wire on the tripod so that it's about 10" away from the breakout point.

11f. Double-check all the wiring and prepare to activate the coil. Plug the NST into your momentary switch, and your momentary switch into mains power.

11g. Push the momentary switch to activate the coil for a few seconds. You should certainly see arcs in the spark gap and will hopefully see an arc from the top-load.

11h. Shut off the coil and disconnect power.

11i. Discharge the six-pack capacitor with the discharge tool.

11j. If you did not see an arc from the top-load, move the ground wire on the tripod a bit closer to the breakout point. Repeat from Step 11f.

11k. Move the alligator clip tap one turn inward on the primary coil. Move the ground wire on

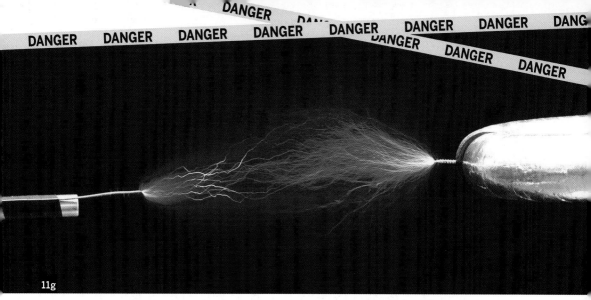

11g

the tripod a bit further from the breakout point. Repeat from Step 11f until you've found the optimal turn for attaching the clip lead.

11l. Repeat Steps 11f–11k to determine the optimal tap point on that turn of the coil for attaching the clip lead, moving the clip from one bare tap point to the next.

11m. Adjust the multispark connections to increase the number of active gaps by 1, and repeat Steps 11f –11i. Repeat this step until the gap stops firing, then back up one position.

Going Further

This Tesla coil has been designed for easy construction from common materials. There are a number of improvements that might enhance performance.

⚡ A more robust dip electrode can be made using glass bottles with screw-on plastic caps. Each bottle's electrode consists of a carriage bolt inserted head-down, with the threads protruding through a hole drilled in the plastic bottle cap, to which it is secured with nuts and washers. Individual bottle electrodes can then be interconnected using jumpers terminated with ring-tongue lugs.

⚡ An air-quenched multispark featuring a fan, blower, or suction system would help keep the spark gap clear of ionized gases that, otherwise, tend to accumulate and impede performance.

⚡ A larger-diameter secondary coil would couple more efficiently with the magnetic field of the primary coil. Similarly, a conical primary coil would improve coupling as well as reduce flash-over to the secondary.

⚡ A more powerful high-voltage supply transformer, whether an NST or otherwise, could of course be made into a more powerful coil.

⚡ Commercial capacitors suitable for off-the-shelf use in a spark gap Tesla coil are rare and expensive. A multiple mini capacitor (MMC) is an excellent DIY alternative. An MMC consists of many smaller, high-quality, high-voltage, off-the-shelf capacitors wired in a series/parallel arrangement to reach the necessary tank capacitance. MMCs are more durable and perform better than bottle capacitors. ◪

Craig Newswanger is a behind-the-scenes member of ArcAttack, the musical Tesla coil crew. He has been an Army photographer, a Disney Imagineer, and a maker of laser light shows and holograms. He built his first computer from a kit in 1975; today he builds all kinds of things at his Resonance Studio Workshop in Austin, Texas.

➕ See the whole project online at makezine.com/35. For more on Tesla coils, check out these great sites:
Bart Anderson — classictesla.com
How Tesla Coils Work — hvtesla.com
Steve Ward — stevehv.4hv.org
Arc Attack — arcattack.com
Terry Fritz Archive Mirror — capturedlightning.org

Brushes with Disaster

Nate Van Dyke

Readers share their experiments gone awry.

DON'T MAKE SMOKE BOMBS INSIDE

By Mark Frauenfelder

When I was 12 years old, I got my hands on a copy of Abbie Hoffman's *Steal This Book*, a guidebook for pranks, protests, and overthrowing the system. My favorite part of the book was the "People's Chemistry" chapter, which had recipes for making stink bombs, smoke bombs, Molotov cocktails, Sterno bombs, aerosol bombs, pipe bombs, and more.

Here's Hoffman's recipe for smoke bombs:

Sometimes it becomes strategically correct to confuse the opposition and provide a smoke screen to aid an escape. A real home-made smoke bomb can be made by combining four parts sugar to six parts saltpeter (available at all chemical supply stores). This mixture must then be heated over a very low flame. It will blend into a plastic substance. When this starts to gel, remove from the heat and allow the plastic to cool. Embed a few wooden match heads into the mass while it's still pliable and attach a fuse.

The smoke bomb itself is non-explosive and non-flame-producing, so no extreme safety requirements are needed. About a pound of the plastic will produce thick enough smoke to fill a city block. Just make sure you know which way the wind is blowing, Weathermen-women!

After a bit of research I learned that the chemical name for saltpeter is potassium nitrate. I called the pharmacy department at the local Kmart and they told me they had it in stock. I rode my bike over and bought a pound. When I got home I set up my Boy Scout camping stove in the backyard and melted a batch of potassium nitrate and sugar in a frying pan. After letting it cool and forming it into a lump, I placed it on the asphalt of the cul-de-sac I lived in and lit it with the match. An astonishing amount of opaque white smoke gushed out of the bomb. The smoke bomb itself liquefied and bubbled like molten lava for quite a while afterward, causing damage to the asphalt.

I invited my friends over and we made more. The next day we bought every container of potassium nitrate that Kmart had in stock. The guy at the pharmacy counter gave us a funny look but didn't ask us why we were buying it.

With all that potassium nitrate we wanted to make a big batch so we'd have plenty of smoke bombs on hand for the Fourth of July. The small camp stove I was using would not be sufficient for our ambitious plans. I asked my friend Garrick if we could cook it up in his kitchen. He thought it was a great idea. The two of us went to his house, dumped a huge amount of potassium nitrate and sugar into an 8-quart pot and cranked up the burner on the range. His mother asked us what we were doing, and we explained that we were making smoke bombs. She asked if it was safe, and I told her that we had made it a bunch of times before and no one had been hurt (which was true). Potassium nitrate is white like sugar, and it looks innocuous, so she just told us to be careful and left the kitchen.

My friend and I took turns stirring the mixture with a wooden spoon, waiting for it to start to get soft and melt. But nothing was happening. We turned up the burner. Ten minutes later, still nothing. Suddenly, my friend remembered that he had the new Ten Years After record and asked if I wanted to listen to it in his room. "Sure," I said. "We can check on the smoke bombs in about 10 minutes."

We were upstairs for less than two minutes when we heard a strange rumbling sound, kind of like a vacuum cleaner but lower pitched and throatier. We looked at each other with fear and ran out of his room and down the stairs. We heard my friend's father scream "Holy sh*t!" By the time we got to the bottom of the stairs the entire first floor was filled with opaque white smoke. I felt like I was lost inside a giant bucket of milk. In the kitchen, we could see puddles of molten smoke bomb mixture on the floor, counters, even dripping off the ceiling. The puddles were glowing red, barely visible through the smoke. My friend's dad had run to the backyard to get the garden hose and was spraying water into the pot and into the burning puddles of smoke bomb lava.

Fortunately, no one was hurt. The property damage wasn't too bad, either, but I got grounded nonetheless. I was especially thankful that my parents didn't forbid me from making smoke bombs. The only stipulation was that I had to go back to making it in our backyard with a camping stove.
—Mark Frauenfelder, Los Angeles

HEADS MAKE BAD BRAKES

I was moving a bookshelf on a brick feature wall just after buying my home. This involved drilling wooden dowels into the brickwork to hold the bookshelf using my new rotary hammer drill. The problem was that I was using a 12mm drill bit and tried to drill between two bricks that had 10mm spacing. I was standing on a chair so I had the height, and started using the drill. Unfortunately, the drill caught and rotated about 270°. What stopped it was my forehead.

As you can imagine, a 1,000-watt drill has a bit of weight behind it, and hit me with some force. Thankfully I was not knocked off the chair, but I was left with a lot of blood and swelling. I decided to stay up most of the night in case I started getting a headache, and when I saw the doctor the next day I ended up getting 11 stitches.
—Darryl Smith, Sydney

INDOOR QUADCOPTER CALAMITIES

I've taken a quadrotor to the arm and still

have the scar. I've over-RPMed carbon fiber blades which got thrown *through* the drywall. Unintentionally destroyed thousands of dollars in electronics for various reasons (normally incompetence). I've also had a quadrotor go to maximum power indoors and smash into the ceiling so hard that the motor shafts left a perfect square pattern of holes about an inch deep.
—Ryan Turner, Calgary, Alberta

SCULPTURE GETS ITS REVENGE

This happened about 43 years ago. I was welding on the base of an 8'-tall sculpture with an oxy-acetylene torch when the sculpture fell over on me. The torch was knocked out of my hand, flipped over, and the flame made a quick pass across the back of my hand, leaving charred flesh in its wake. Very painful, but I recovered with no visible damage.
—Tom Dimock, Ithaca, N.Y.

BOWL GOES BANG

Back when I was young and stupid (I've since graduated to much more mature mistakes), I was casting some small resin parts using silicone molds. Some had a few bubbles, and I'd read somewhere that a vacuum chamber would draw the air out of the resin. I cobbled together a vacuum chamber. I took a plastic cereal bowl, cut a hole in the bottom, stuck a plastic hose through it, and sealed it with some sort of caulk. I poured resin into the mold and placed the bowl over it. I flipped the switch on the little vacuum pump and waited.

Have you ever heard the huge boom that an imploding cereal bowl makes when it shatters into dozens of sharp pieces? Thankfully I was living in suburban Ohio at the time, where big booms make up the soundtrack of a normal day.
—Dave Barak, San Diego

TOWER OF FAIL

I live in Seattle. Every year we have unlimited hydroplane races on Lake Washington. The place to watch the action from was out on the lake. The problem was, unless you paid to get on the front row of boats, you couldn't see anything. No problem — we built a series of bigger and bigger floating platforms. Our final one was a pyramid of aluminum irrigation pipe designed to hold 10 people 18' off the water, with a seat on top 26' off the water. It floated on foam blocks.

Problem — traffic control! In our earlier days we used a rope ladder. If it got too crowded we'd just pull up the ladder. Later on we got lazy and just hung an aluminum ladder overboard; it was much easier to climb but a real hassle to pull up so we just left it down. That year we became victims of our own popularity — too many people climbed up, sinking one of the floats and making the whole thing — with 12 people on it — topple over in slow motion. Fortunately nobody got hurt, but the Coast Guard came to see if everybody was OK and after that, the boat race people changed the rules and outlawed our tower! Now only one row of registered floating vessels can be out on the water watching the races.
—Paul Shemeta, Seattle

SHOCK BOX

I had an old AT power supply and wanted to use it as a bench power supply, but it only produced 11.5V, not 12. So I had the brilliant idea of opening it up and seeing if there was anything I could fiddle with to get an extra .5V.

Turns out that the fan, power connectors, and a couple of other components were glued to one half of the case, while the main board was resting on the other side, so it ended up sitting on the desk in a heap while I used my plastic-handled jeweler's screwdrivers (big chunk of plastic between me and the hot electric death? Safe!) to twiddle the trimpots. After some fiddling, I got it up to 11.8V.

Then I decided to move it 10cm along the desk so I could reach something else. Grabbed both halves of the metal case in my hands and lifted it slightly to slide it across the desk. I got the biggest and most dangerous shock I've ever experienced, dropped it, yanked out the power cable, turned the switch off at the wall

and swore off ever working on anything that may have mains power in it unless there was someone else in the house.
—Julian Calaby, Melbourne, Australia

EXPLODING EASTER EGG

I once decided to streamline the Easter egg dying process by hard-boiling and dying an egg in one step in my mother's microwave oven.

Steps in recreating my experiment:
1. Place egg and dying liquids in ceramic mug.
2. Place mug in Mom's microwave.
3. Set power to high for several minutes.
4. Hear door blow off of microwave oven.
5. Clean purple egg off the kitchen ceiling.
6. Apologize profusely.

Oh yeah, I was approximately 30 years old when I attempted this feat of genius.
—Gary Stager, Torrance, Calif.

IF YOU TRIED THIS NOW YOU'D BE IN JAIL

My high school chemistry teacher let a few of us do extracurricular experiments in the lab unsupervised after class (try that now!), so it was only natural that we would whip up a bit of nitrogen triiodide, a weak explosive that is so unstable when dry that it can be set off if a fly lands on it. We made a mound of the stuff and whacked it with a stick as a test. It went up like a cloud, showering the room with bits of explosive. When we saw the mini mushroom, we knew it was time to get out of there and pretend nothing had happened. When I closed my open chemistry book, the inside pages exploded. The next day anyone walking into the room would hear popping sounds under their feet or when they sat down. We knew the gig was up. It took two hours of decontamination with a wet mop and sponges, under the teacher's watchful eye this time.
—Matthew Gryczan, Grand Rapids, Mich. ◪

DANGER DANG

DANGER DANGER DANGER DANGER DANGER DANGER DANGER DANGER

◿ **TIME: AN AFTERNOON** ◿ **COST: $10**

Homemade Sugar Rocket

Written by
William Gurstelle
Photographed by
Gunther Kirsch

Cook up a solid-fuel rocket engine and let it fly.

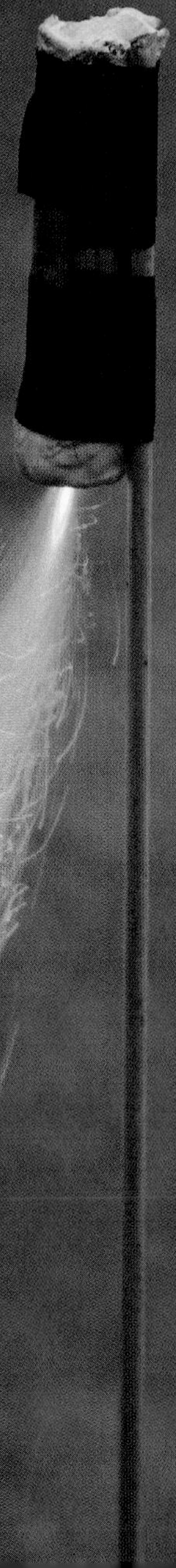

⚠ Undertake this project at your own risk. You are literally playing with fire, so understand what you're getting into and don't sell the dangerous aspects of this activity short. Sugar rocket fuel burns fiercely — do not ignite it until it's contained in a completed motor. Follow all instructions and safety precautions carefully.

Many hobby stores sell model rocketry supplies, but I think making your own rocket engine from scratch is a more meaningful and exciting experience. In this project, we'll build on the work of rocket pioneer Jack Parsons (see MAKE Volume 13, "Darkside Rocketeer," for more about this fascinating character). It was Parsons who invented "castable" rocket fuel, which starts as a soft, pliable material and slowly hardens, allowing it to be cast or molded into a high-performance motor.

In this project, you'll combine two commonly available substances — granulated sugar and potassium nitrate — to make a powerful engine that can propel a small rocket to impressive heights. This method involves melting a mixture of sugary fuel and chemical oxidizer (the potassium nitrate) over an electric hot plate and then pouring it into a paper rocket body where it solidifies into a rock-hard casting containing an incredible amount of chemical energy.

William Gurstelle

1

MATERIALS

» **Kraft paper** Brown grocery bags work fine
» **Wooden dowel, ³⁄₈" diameter**
» **White glue, all-purpose**
» **Water putty, nonshrinking** from a
    hardware store
» **Sugar, white, 7g**
» **Potassium nitrate aka saltpeter (KNO₃), 14g**
    Some rocket enthusiasts use saltpeter-based
    stump remover, but I've never had good luck
    with it. Better to buy the real McCoy online.
    It's cheap and the quality is terrific. Buy the
    powdered, not the "prilled" stuff.
» **Nail, 6D** aka 6-penny
» **Thin wooden stick** such as a bamboo
    shish-kebab skewer
» **Visco fuse** aka safety fuse or cannon fuse.
    Search the internet for vendors.
FOR LAUNCH PLATFORM:
» **Block of wood,** scrap
» **Metal pipe, ¾"- 1"-diameter, 8"-long**
» **Floor flange**
» **Wood screws**

TOOLS

» **Safety glasses**
» **Lab apron**
» **Fire extinguisher, Type A**
» **Scissors**
» **Ruler**
» **Fine sandpaper**
» **Electric drill with ⁷⁄₆₄" bit**
» **Scale** that can measure grams
» **Plastic mixing container with a
    tight-fitting lid**
» **Lead fishing weights, large; or lead balls,
    .50 caliber (12)** available at hunting and
    fishing stores. *Do not* substitute steel balls
    or glass marbles.
» **Electric skillet, or electric hot plate with
    heavy pan** *Do not* use a gas hot plate or burner,
    or any kind of open flame.
» **Spatula, plastic**
» **Mallet, rubber**
» **Bucket of water** in case of a dud
» **Heavy gloves** in case of a dud

1. Wind the motor casing.

Cut a sheet of kraft paper 4"×10". Place the
³⁄₈" dowel on one of the 4" edges, then wrap
the paper tightly around the dowel. As you
roll, apply a very thin coat of white glue over
one whole side of the paper. Roll the paper
tube as tightly as possible. Remove the dowel
and let the tube dry.

After the glue dries, cut the paper tube
into two 2"-long segments. Each tube will be
suitable for a single-use motor casing.

2. Make the rocket nozzle.

Reinsert the ³⁄₈" dowel into the casing,
leaving a ⁵⁄₁₆" gap between the end of the
tube and the dowel. Mix a small amount of
nonshrinking water putty according to the
label directions. Press the putty firmly into
the end of the tube. Slowly remove the dowel
and set the nozzle aside to dry. Repeat this
process for the second motor.

When the nozzle is dry, use fine sandpaper
to smooth and flatten its exposed surface.
Mark the center of the nozzle and drill a
⁵⁄₃₂"-diameter hole through the center.

3. Prepare the rocket fuel.

Using the scale, measure
14 grams of potassium
nitrate or saltpeter (KNO₃)
and 7 grams of sugar. This
makes enough propellant for
a single 2"-long rocket motor.

CAUTION:
Put on your safety
glasses and leave them
on for the rest of
the project.

Blend the ingredients by placing them
into a tightly covered plastic container with
a handful of large lead fishing weights or

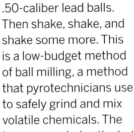

.50-caliber lead balls. Then shake, shake, and shake some more. This is a low-budget method of ball milling, a method that pyrotechnicians use to safely grind and mix volatile chemicals. The longer you shake, the better the mixing and the better the rocket fuel.

CAUTION: You're shaking a container of rocket fuel, so use only lead balls, as other mixing media can cause sparks. Keep the cover on tight, and keep the fuel far from ignition sources.

Do not use a gas burner or any open flame!

Now go outdoors to do the next part. Heat an electric skillet or a heavy pan on a carefully controlled electric hot plate to 285°F. (You might need an extension cord.)

Add the rocket-fuel mix to the pan. Stir continually with a hard plastic spatula or spoon. The mixture will soon melt into a peanut butter-like consistency. After a few more minutes of heating and stirring, it will melt into a smooth paste. Turn off the heat. The mixture will stay plastic and moldable just long enough for you to make the rocket motor.

4. Load the fuel.

Let the rocket-fuel mixture cool until you can just handle it with your fingers, and then immediately stuff the hot, pliable fuel into the motor casing. Insert the 3/8" dowel into the open end of the motor and gently tap it with the rubber mallet until you have evenly filled the motor up to 3/8" from the top.

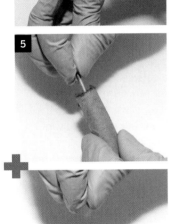

5. Core the rocket engine.

To produce enough thrust to lift your rocket into the air, you'll need to make what rocketeers refer to as a *core burner* (as opposed to an *end burner*). This style of rocket motor has an open center, increasing the surface area over which the rocket fuel burns.

While the fuel is still hot, insert the 6D nail into the nozzle and push it through the fuel, twisting it as you go, until the tip emerges on the other end. Carefully remove the nail by rotating it slowly so that the cylindrical hole remains intact.

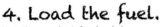

6. Seal the open end.

Fill the 3/8" gap at the open end of the motor with water putty. Let it dry completely.

7. Attach balance sticks.

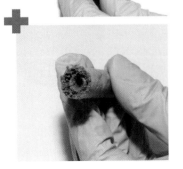

Glue a shish-kebab skewer or very narrow wooden dowel to the rocket casing, so it extends backward from the nozzle. The length depends on the weight of the rocket: choose a length such that the rocket body and stick will balance horizontally at

a point 1" behind the nozzle.

Set your rocket aside in a safe and secure location. When the glue and putty have dried, it's complete. It should be constructed as you see in the cutaway diagram. The fuel will absorb moisture from the air, so use the rocket as soon as possible.

8. Build a launch platform.

Make a simple launching platform by attaching an 8" length of ¾"- or 1"-diameter metal pipe to a scrap block of wood, using a floor flange and wood screws to support the pipe.

9. Launch your sugar rocket!

Your rocket could reach altitudes of several hundred feet. Since a rocket's trajectory is difficult to predict, you must launch only in appropriate areas in appropriate seasons (for example, flat, open, and free from easily ignitable materials). Check local laws and make sure you have the landowner's permission to launch.

Insert a 2"-long fuse directly into the rocket's nozzle. If necessary, bend the end of the fuse before inserting it, so it will stay in the nozzle. Light the fuse and immediately retreat to a safe distance. Blastoff!

Troubleshooting

If the rocket does not ignite, stay away from it for 5 minutes. Then put on thick gloves, pick it up carefully with the nozzle pointed away, and soak it in water until it disintegrates. Discard the pieces in an outdoor trash can.

Be certain your chemicals are pure and well mixed. When in doubt, increase the mixing time in the shaking container.

Pay careful attention to the temperature setting on the electric skillet or hot plate. Melting at the wrong temperature can hurt the fuel's performance.

If your rocket fires but won't lift off, increase the size of the nozzle opening, increase the size of the core hole, or reduce the rocket's weight by using thinner paper and less glue in building the tube. ◪

7

ROCKET CUTAWAY

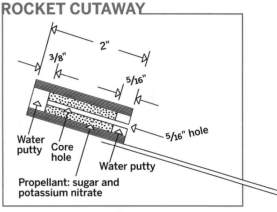

2"

3/8"

5/16"

5/16" hole

Water putty

Core hole

Water putty

Propellant: sugar and potassium nitrate

8

9

The new and expanded edition of William Gurstelle's book Backyard Ballistics is now available in the Maker Shed (makershed.com).

⚡ **TIME: AN AFTERNOON**   ⚡ **COST: $600**

High-Power Sugar Rocket

Written by
Dan Pollino

The humble beginnings of a wicked DIY rocket.

WARNING:
Goes *really high*. Yes, you will need FAA clearance to launch this rocket!

Dan Pollino's 6-foot sugar rocket accelerates to 400mph in 3 seconds and goes 6,000 feet high. It's an awesome DIY project. To build it, get Dan's excellent how-to book *I Still Have All My Fingers: How to Build a Big Sugar Rocket on a Budget Without Losing a Limb*.

We asked Dan how he got into rockets. He took us all the way back.

When I was in sixth grade, I had an unconventional teacher named Mr. Zimmerman. He taught us fractions by giving us fake seed money to buy and sell stocks every day. He set up a competition to see who could grow their seed money the most in a month. That was the most exciting math lesson I ever had and I didn't even realize I was learning it.

In science, he used a series of experiments to teach us the principles of physics. There was the typical egg drop, where we designed a container to prevent an egg from breaking when it was dropped three stories onto a parking lot. But the one everyone looked forward to all year was the model rocket launch. Of course there were lectures on the principles of rocket flight, but the anticipation of the launch made the laws of the physical universe seem like rules to a game and not a boring lecture.

Everyone had to build their own rocket and bring it to class on the day of the launch. Back in the day, Estes model rockets were available everywhere and there wasn't any problem bringing a model rocket engine, a Class 1.4 explosive, to school.

My parents could only afford the engine, so I had to make the rocket myself. No X-wing fighter or Atlas model for me. Mine was constructed from a discarded paper towel tube. It had cardboard fins that were glued on — poorly— and a paper nose cone. I made the

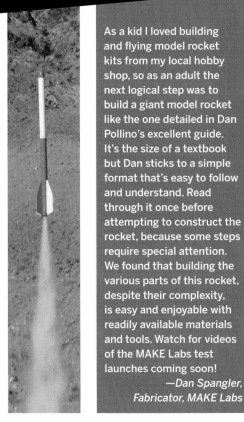

As a kid I loved building and flying model rocket kits from my local hobby shop, so as an adult the next logical step was to build a giant model rocket like the one detailed in Dan Pollino's excellent guide. It's the size of a textbook but Dan sticks to a simple format that's easy to follow and understand. Read through it once before attempting to construct the rocket, because some steps require special attention. We found that building the various parts of this rocket, despite their complexity, is easy and enjoyable with readily available materials and tools. Watch for videos of the MAKE Labs test launches coming soon!
—Dan Spangler, Fabricator, MAKE Labs

parachute from a plastic garbage bag and used shoelaces for the shroud lines. Needless to say it weighed a ton but I was proud of it nonetheless.

Launch day came and the entire class went to the track field behind the school to get our final grade. I put my rocket on the pad, we all did a countdown, and up it went ... all of about 2 feet. The rocket just hung there for a second, fell back down on its side with a thud, and after a second or two, the parachute puffed out.

Everyone laughed and I got a horrible grade, but none of that was important to me. What I did learn, and what turned out to be an invaluable lesson, was that I could make something. ◪

Dan Pollino is the author of several how-to books. His rockets have been featured on G4 TV's It's Effin' Science. His website inverseengineering.com focuses on amateur rocketry in California.

Inside the Six-Foot Sugar Rocket

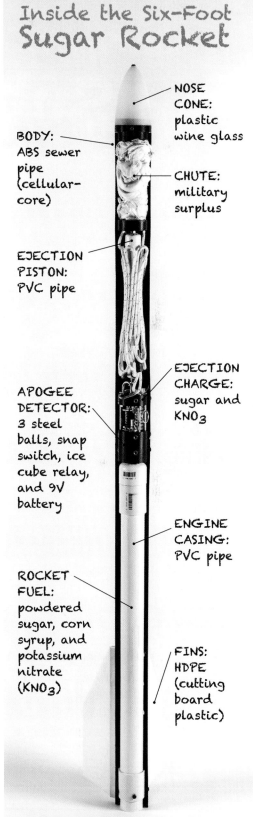

NOSE CONE: plastic wine glass

BODY: ABS sewer pipe (cellular-core)

CHUTE: military surplus

EJECTION PISTON: PVC pipe

EJECTION CHARGE: sugar and KNO_3

APOGEE DETECTOR: 3 steel balls, snap switch, ice cube relay, and 9V battery

ENGINE CASING: PVC pipe

ROCKET FUEL: powdered sugar, corn syrup, and potassium nitrate (KNO_3)

FINS: HDPE (cutting board plastic)

DANGER DANGER DANGER DANGER DANGER DANGER DANGER DANGER DANGER DANGER DANGER DANGER DANGER DANGER DANGER DANGER

TIME: 2–3 HOURS   **COST: $60–$90**

The Sublimator Dry Ice Cannon

Unleash carbon dioxide gas to launch T-shirts and other projectiles.

Written by William Gurstelle

⚠ DRY ICE SAFETY RULES

1. Do not store dry ice in an airtight container. The best place to store dry ice is in a styrofoam cooler with a loose-fitting lid.
2. Do not touch dry ice with your bare skin; you could get frostbite. Always use insulated gloves or tongs.
3. Open the windows in your car or workshop when you transport or experiment with dry ice. Breathing high concentrations of carbon dioxide is hazardous and can be deadly.
4. Use only the amount of dry ice specified in the directions.

Is this project a bit dangerous? Sure. But with care and planning you can safely build a wicked-cool dry ice cannon that uses the power of sublimation to shoot a projectile high into the sky.

Frozen carbon dioxide is called "dry ice" because it doesn't melt into liquid like normal ice. Water, as we all know, becomes solid ice below 32°F, exists as a liquid between 32°F and 212°F, and turns to steam at temperatures above that. Dry ice is different. There's not enough pressure in our atmosphere for carbon dioxide to exist as a liquid. (On Venus, there are oceans of liquid CO_2 because the atmospheric pressure is so high.) At an Earthly pressure of 14.7psi, dry ice

DANGER DANGER DANGER DANGER DANGER DANGER

never melts to a liquid state. Instead, at −109°F, it passes directly from a solid to a gas. That's a physics phenomenon called *sublimation*.

It's sublimation that makes this dry ice cannon shoot. As dry ice sublimates (or sublimes), the CO_2 gas produced rapidly builds up pressure inside the cannon. To control the release of the pressure, we'll install a *burst valve* that will suddenly but safely open when the pressure reaches a predetermined threshold.

Our burst valve is a disc of aluminum foil held in place between the 2 sections of a PVC pipe union and sealed tightly by the O-ring built into the union. When the pressure inside the cannon reaches about 25psi (far below the rupture strength of the PVC), the foil membrane bursts, the pressure escapes to the barrel, and whatever's in the barrel is launched high into the air.

Build Your Sublimator Cannon

Attempt this project at your own risk. Follow the directions carefully and be aware that material flaws and poorly constructed joints can lead to big trouble.

1. Cut the pipe to length, referring to the assembly drawing. Before cementing, remove all PVC shavings with a cloth or water to get a clean bond (and to avoid painful high-speed PVC splinters upon firing).

For tips on cutting and gluing PVC, visit makezine.com/35.

2. Cement the 4" end cap to the 4" pipe. Cement the 4"–2" reducer fitting to the other end of the 4" pipe. This is your pressure chamber.

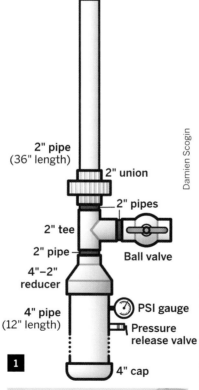

2" pipe (36" length)

2" union

2" pipes

2" tee

Ball valve

2" pipe

4"–2" reducer

4" pipe (12" length)

PSI gauge

Pressure release valve

1

4" cap

Damien Scogin

2

Gunther Kirsch

MATERIALS

» **PVC pipe, Schedule 40:** 4" diameter, 12" length; 2" diameter, 3' length (1) and 3" lengths (3)
» **PVC pipe fittings, Schedule 40, slip fit:** 4" end cap; 4" to 2" reducer; 2" tee; 2" union with O-ring; 2" ball valve A bell-shaped reducer looks great but is harder to find; look online.
» **Wood dowel or broom handle,** 4' length
» **Pressure gauge, 0–60psi,** with ¼" NPT connector
» **Pressure-release valve, 60psi,** with ¼" NPT connector
» **PVC cement** We recommend using a 2-part PVC primer/cement system for strength.
» **Pipe thread sealant**
» **Duct tape**
» **Dry ice pellets, ½" diameter, 2" long** Industrial gas vendors such as Airgas sell these for about 50¢ a pound.
» **Warm water**
» **T-shirt or towel** rolled and taped so it fits into the barrel
» **Aluminum foil, heavy duty**

FOR EXTRA 2½" BARREL (OPTIONAL):
» **PVC pipe, Schedule 40:** 2½" diameter, 3' length; 2" diameter, 3" length
» **PVC pipe fittings, Schedule 40, slip fit:** 2" union, 2½" to 2" reducer, 2½" coupler

TOOLS

» **Hacksaw**
» **Drill and** $^7/_{16}$" **bit**
» **Tap wrench and tap, ¼" NPT**
» **Pipe wrench**
» **Safety glasses**
» **Gloves, insulated**
» **Tongs (optional)**

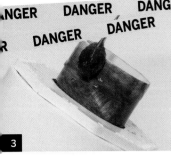

3

4

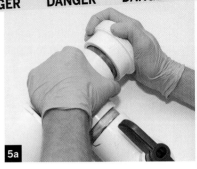

5a

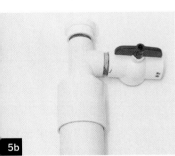

5b

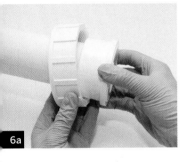

6a

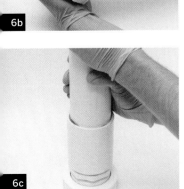

6b

6c

3. Connect the reducer to the tee fitting by cementing each to a 3" length of 2"-diameter pipe.

4. Cement another 3" length of 2" pipe into the middle opening of the tee, then cement the ball valve to this short pipe.

5. Connect the threaded half of the pipe union (the side with the O-ring) to the remaining opening of the tee by using the third 3" length of 2" pipe.

6. Attach the 3'-long 2" pipe to the remaining half of the pipe union. This is the barrel (**Figure 6a**).

OPTIONAL: A bigger, 2½" barrel is perfect for launching tennis balls, though it won't shoot T-shirts as far. To make one, cement a 3" length of 2" pipe into the pipe union (buy a spare union if you want swappable barrels), and on the other end cement the 2½" to 2" reducer, then the 2½" coupling, then the 3' length of 2½" pipe. Be sure to slide the threaded collar onto the pipe union before gluing on the reducer (**Figures 6b and 6c**).

7. Wrap the pressure chamber in 3 layers of your favorite duct tape for an extra margin of safety in case it should crack.

8. Drill a ⁷⁄₁₆" hole in the pressure chamber, near the top of the 4" pipe. Tap threads in the hole for a ¼ NPT pipe. Apply plenty of thread sealant to the pressure gauge's threads and screw it into the hole.

9. Drill a second ⁷⁄₁₆" hole in the pressure chamber, also near the top. Tap the hole for a ¼" NPT pipe. Apply thread sealant to the pressure-release valve and screw it in. Don't mount this valve near the bottom of the chamber, or it'll blow water instead of air.

10. Rinse all PVC shavings out of the cannon with plenty of water.

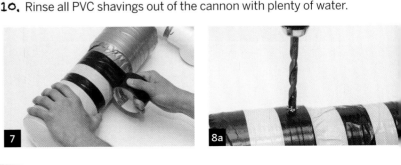

7

8a

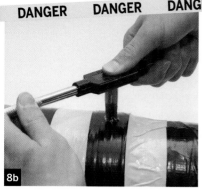

8b

8c

9

Operating the Sublimator Cannon

WARNING: Do not hold the cannon while you fire it!

1. Roll up a T-shirt or towel to fit the barrel, and tape it to hold its shape. Use the dowel or broomstick to push it down the barrel (**Figure A**).

2. Cut a 2½" disc of heavy-duty aluminum foil. Unscrew the pipe union and place the disc between the 2 halves of the union (**Figure B**). Screw the union together tightly. The O-ring must seal the burst disc in place with no air leakage. Two discs can be used for greater range, but start out with one.

3. Open the ball valve and pour 3 or 4 pints of warm water into the chamber (**Figure C**).

4. Secure the cannon so that it's aimed straight up or nearly so.

5. Put on safety glasses and insulated gloves. Push 3 dry ice pellets through the valve, making sure they drop all the way into the water inside the pressure chamber (**Figure D**). Use the broomstick to push them if necessary. Quickly close the valve completely.

6. Immediately, the dry ice will begin to boil inside the Sublimator. Step back a safe distance, and keep an eye on the pressure gauge. The sublimating dry ice causes the pressure to rise until it exceeds the ability of the burst disc to contain it. The disc bursts open, allowing the CO_2 to escape all at once — *pop!* — launching the T-shirt out of the barrel and high into the air!

The new, expanded edition of William Gurstelle's book Backyard Ballistics is now available in the Maker Shed (makershed.com).

A

B

C

D

KEEPING SAFETY IN MIND

- This cannon packs a wallop! Always aim it away from people and things you don't want shot.
- Use caution when handling dry ice. Use only the amount of dry ice recommended.
- The time between loading the cannon and actual firing is variable, depending on water temperature, the amount of dry ice, and the amount of water used. Be patient and never, ever look down the barrel.
- If you have a misfire where the pressure of the sublimating dry ice doesn't build sufficiently to burst the aluminum discs, carefully open the ball valve to release the pressure. Clear the area when doing this, and stand off to the side as you work.
- This device is designed to shoot T-shirts, towels, or other soft objects only.

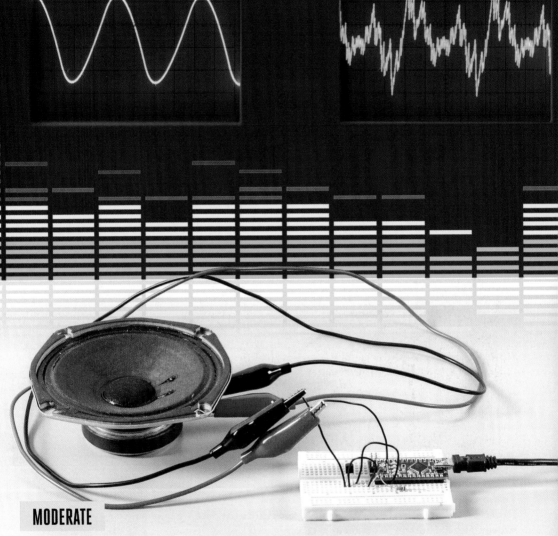

MODERATE

ADVANCED ARDUINO SOUND SYNTHESIS

From "bit banging" to morphing and fading.

Written by **Jon Thompson**

THE ARDUINO IS AN AMAZING PLATFORM for all kinds of projects, but when it comes to generating sound, many users struggle to get beyond simple beeps. With a deeper understanding of the hardware, you can use Arduino to generate any waveform you can imagine, and manipulate it in real time.

Basic Sound Output

"Bit banging" is the most basic method of producing sound from an Arduino. Just connect a digital output pin to a small speaker and then rapidly and repeatedly flip the pin between high and low. This is how the Arduino's `tone()` statement works. The output pins can even drive a small (4cm or less) 8-ohm speaker connected directly between the pin and ground without any amplification.

Cycling the pin once from low to high and back again creates a single square wave (**Figure A**). Time spent in the high state is called *mark time* and time spent low, *space time*. Varying the ratio between mark and space times, aka the *duty cycle*, without changing the frequency of the wave, will change the quality or "timbre" of the sound.

The Arduino's `analogWrite()` function, which outputs a square wave at a fixed frequency of 490Hz, is handy to illustrate the concept. Connect your speaker to pin D9 and ground (**Figure B**) and run this sketch:

```
void setup() {
   pinMode(9,OUTPUT);
}

void loop() {
   for (int i=0; i<255; i++) {
     analogWrite(9,i);
   delay(10);
   }
}
```

You should hear a tone of constant pitch, with a timbre slowly changing from thin and reedy (mostly space time) to round and fluty (equal mark and space time), and back to thin and reedy again (mostly mark time).

A square wave with a variable duty cycle

Gunther Kirsch

TOOLS & MATERIALS
» **Arduino Nano v3.0 microcontroller board**
» **Solderless breadboard**
» **Capacitor, 0.1μF ceramic**
» **Resistor, 10kΩ**
» **Speaker, 8Ω, approx. 4cm diameter**
» **NPN transistor, BC548 type** or similar
» **LED (optional)**
» **Computer running Arduino IDE software** free from arduino.cc
» **Mini-B USB cable**
» **Oscilloscope** If you don't have access to a hardware oscilloscope, check out Christian Zeitnitz's Soundcard Scope software.

» **Code listings 1–6** Each is a complete running Arduino sketch. Download them from makezine.com/35.

EXTRA RESOURCES:
» **OCR2A frequency table** from makezine.com/35
» **Wave table spreadsheet** from makezine.com/35
» **Soundcard Scope software:** zeitnitz.de/Christian/scope_en

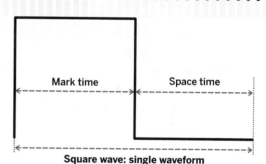

Mark time   Space time

Square wave: single waveform

A

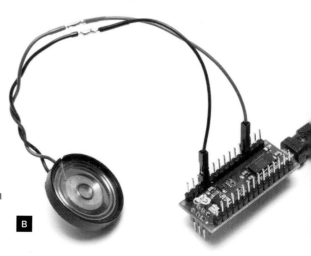

B

is properly called a *pulse-width modu-lated (PWM)* wave. Altering the duty cycle to change timbre may serve very basic sound functions, but to produce more complex output, you'll need a more advanced approach.

From Digital to Analog

PWM waves are strictly digital, either high or low. For analog waves, we need to generate voltage levels that lie between these 2 extremes. This is the job of a *digital-to-analog converter (DAC)*.

There are several types of DAC. The simplest is probably the R-2R ladder (**Figure C**). In this example, we have 4 digital inputs, marked D0–D3. D0 is the least significant bit and D3 the most significant.

If you set D0 high, its current has to pass through a large resistance of $2R + R + R + R = 5R$ to reach the output. Some of the current also leaks to ground through the relatively small resistance $2R$. Thus a high voltage at D0 produces a much smaller output voltage than a high voltage at D3, which faces a small resistance of only $2R$ to reach the output, and a large resistance of $5R$ to leak to ground.

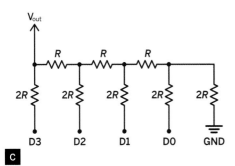

C

Setting D0–D3 to binary values from `0000` to `1111` (0–15 decimal), and then back down to `000` in quick succession, ought to output the triangle wave shown in **Figure D**. To produce other waveforms, in theory, we must simply present the right sequence of binary numbers to D0–D3 at the right rate.

Unfortunately, there are drawbacks to using an R-2R DAC, foremost probably that it requires very precise resistor values to prevent compound errors from adding up and distorting the waveform. The jagged "steps" must

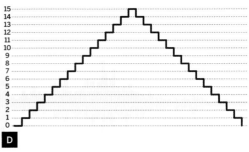

D

also be smoothed, using a low-pass filter, to prevent a discordant metallic sound. Finally, an R-2R DAC uses up more output pins than are strictly necessary.

Though a bit harder to understand, the "1-bit" DAC produces very smooth, high quality waveforms using just one output pin with a single resistor and capacitor as a filter. It also leaves the Arduino free to do other things while the sound is playing.

One-Bit DAC Theory

If you replace the speaker from the bit-banging sketch with an LED, you will see it increase in brightness as the duty cycle increases from 0 to 100%. Between these 2 extremes, the LED is really flashing at around 490Hz, but we see these flashes as a continuous brightness.

This "smoothing" phenomenon is called "persistence of vision," and it can be thought of as a visual analogy to the low-pass filter circuit shown in **Figure E**. You can use this filter to smooth the output from a 1-bit DAC.

The mark time of the incoming PWM wave determines the voltage at V_{out} from moment to moment. For example, a mark/space ratio of 50:50 outputs 50% of the high voltage of the incoming signal, a 75:25 ratio outputs 75%

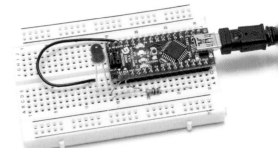

An LED subjected to "bit banging" in place of the speaker will steadily increase in brightness.

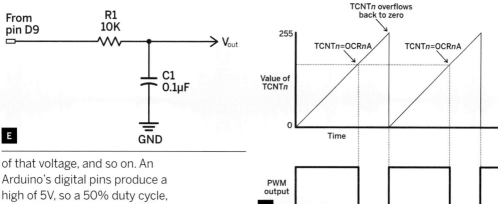

E

F

of that voltage, and so on. An Arduino's digital pins produce a high of 5V, so a 50% duty cycle, for example, would give 2.5V at V_{out}.

For best sound quality, the frequency of the PWM signal should be as high as possible. Luckily, the Arduino can produce fast PWM waves up to 62.5KHz. The hardware also provides a handy mechanism for updating the mark time from a lookup table at absolutely regular intervals, while leaving the Arduino free to do other things.

The Arduino 1-Bit DAC

The ATmega328 chip at the heart of the Arduino Nano 3 contains 3 hardware timers. Each timer includes a counter that increments at each clock tick, automatically overflowing back to 0 at the end of its range. The counters are named TCNT*n*, where *n* is the number of the timer in question.

Timer0 and timer2 are 8-bit timers, so TCNT0 and TCNT2 repeatedly count from 0 to 255. Timer1 is a 16-bit timer, so TCNT1 repeatedly counts from 0 to 65535, and can also be made to work in 8-bit mode. In fact, each timer has a few different modes. The one we need is called "fast PWM," which is only

Listing 1

```
#include <avr/interrupt.h> // Use timer interrupt library

/******** Sine wave parameters ********/
#define PI2     6.283185 // 2*PI saves calculation later
#define AMP     127 // Scaling factor for sine wave
#define OFFSET  128 // Offset shifts wave to all >0 values

/******** Lookup table ********/
#define LENGTH  256  // Length of the wave lookup table
byte wave[LENGTH];  // Storage for waveform

void setup() {

/* Populate the waveform table with a sine wave */
for (int i=0; i<LENGTH; i++) { // Step across wave table
    float v = (AMP*sin((PI2/LENGTH)*i));  // Compute value
    wave[i] = int(v+OFFSET); // Store value as integer
  }

/****Set timer1 for 8-bit fast PWM output ****/
  pinMode(9, OUTPUT);  // Make timer's PWM pin an output
  TCCR1B  = (1 << CS10); // Set prescaler to full 16MHz
  TCCR1A |= (1 << COM1A1);  // Pin low when TCNT1=OCR1A
  TCCR1A |= (1 << WGM10);   // Use 8-bit fast PWM mode
  TCCR1B |= (1 << WGM12);

/******** Set up timer2 to call ISR ********/
  TCCR2A = 0; // No options in control register A
  TCCR2B = (1 << CS21); // Set prescaler to divide by 8
  TIMSK2 = (1 << OCIE2A); // Call ISR when TCNT2 = OCRA2
  OCR2A = 32;  // Set frequency of generated wave
  sei();  // Enable interrupts to generate waveform!
}

void loop() {  // Nothing to do!
}

/******** Called every time TCNT2 = OCR2A ********/
ISR(TIMER2_COMPA_vect) {  // Called when TCNT2 == OCR2A
  static byte index=0;  // Points to each table entry
  OCR1AL = wave[index++]; // Update the PWM output
  asm("NOP;NOP");  // Fine tuning
  TCNT2 = 6;  // Timing to compensate for ISR run time
}
```

available on timer1.

In this mode, whenever TCNT1 overflows to zero, the output goes high to mark the start of the next cycle. To set the mark time, timer1 contains a register called OCR1A. When TCNT1 has counted up to the value stored in OCR1A, the output goes low, ending the cycle's mark time and beginning its space time. TCNT1 keeps on incrementing until it overflows, and the process begins again.

This process is represented graphically in **Figure F,** previous page. The higher we set OCR1A, the longer the mark time of the PWM output, and the higher the voltage at V_{out}. By updating OCR1A at regular intervals from a pre-calculated lookup table, we can generate any waveform we like.

Basic Wave Table Playback

Listing 1 (previous page) contains a sketch that uses a lookup table, fast PWM mode, and a 1-bit DAC to generate a sine wave.

First we calculate the waveform and store it in an array as a series of bytes. These will be loaded directly into OCR1A at the appropriate time. We then start timer1 generating a fast PWM wave. Because timer1 is 16-bit by default, we also have to set it to 8-bit mode.

We use timer2 to regularly interrupt the CPU and call a special function to load OCR1A with the next value in the waveform. This function is called an *interrupt service routine (ISR)*, and is called by timer2 whenever TCNT2 becomes to equal OCR2A. The ISR itself is written just like any other function, except that it has no return type.

The Arduino Nano's system clock runs at 16MHz, which will cause timer2 to call the ISR far too quickly. We must slow it down by engaging the "prescaler" hardware, which divides the frequency of system clock pulses before letting them increment TCNT2. We'll set the prescaler to divide by 8, which makes TCNT2 update at 2MHz.

To control the frequency of the generated waveform, we simply set OCR2A. To calculate the frequency of the resulting wave, divide the rate at which TCNT2 is updated (2MHz) by the value of OCR2A, and divide the result by the

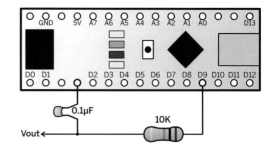

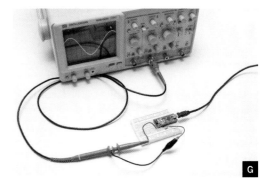

G

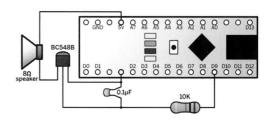

H

length of the lookup table. Setting OCR2A to 128, for example, gives a frequency of:

$$\frac{\text{TCNT2 rate}}{\text{OCR2A value} \times \text{wavetable length}} = \frac{2{,}000{,}000\text{Hz}}{128 \times 256} = 61.04\text{Hz}$$

which is roughly the B that's 2 octaves below middle C. For a table of values giving standard musical notes, see makezine.com/35.

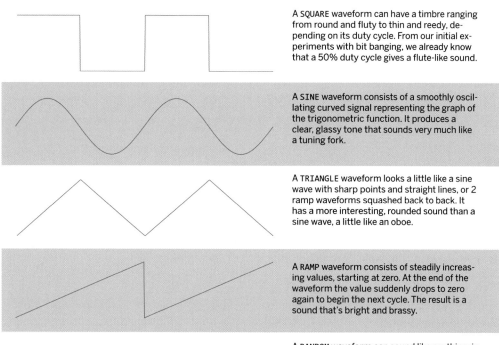

A SQUARE waveform can have a timbre ranging from round and fluty to thin and reedy, depending on its duty cycle. From our initial experiments with bit banging, we already know that a 50% duty cycle gives a flute-like sound.

A SINE waveform consists of a smoothly oscillating curved signal representing the graph of the trigonometric function. It produces a clear, glassy tone that sounds very much like a tuning fork.

A TRIANGLE waveform looks a little like a sine wave with sharp points and straight lines, or 2 ramp waveforms squashed back to back. It has a more interesting, rounded sound than a sine wave, a little like an oboe.

A RAMP waveform consists of steadily increasing values, starting at zero. At the end of the waveform the value suddenly drops to zero again to begin the next cycle. The result is a sound that's bright and brassy.

A RANDOM waveform can sound like anything, in theory, but usually sounds like noisy static. The Arduino produces only pseudorandom numbers, so a particular randomSeed() value always gives the same "random" waveform.

The ISR takes some time to run, for which we compensate by setting TCNT2 to 6, rather than 0, just before returning. To further tighten the timing, I've added the instruction asm("NOP;NOP"), executing 2 "no operation" instructions using one clock cycle each.

Run the sketch and connect a resistor and capacitor (**Figure G**). You should see a smooth sine wave on connecting an oscilloscope to V_out. If you want to hear the output through a small speaker, add a transistor to boost the signal (**Figure H**).

Programming Simple Waves

Once you know how to "play" a wave from a lookup table, creating any sound you want is as easy as storing the right values in the table beforehand. Your only limits are the Arduino's relatively low speed and memory capacity.

Listing 2 (available at makezine.com/35) contains a waveform() function to prepopulate the table with simple waveforms: SQUARE, SINE, TRIANGLE, RAMP, and RANDOM. Play them to see how they sound (**Figure I**).

The RANDOM function just fills the table with pseudorandom integers based on a seed value. Changing the seed value using the randomSeed() function allows us to generate different pseudorandom sequences and see what they sound like. Some sound thin and weedy, others more organic. These random waveforms are interesting but noisy. We need a better way of shaping complex waves.

Additive Synthesis

In the 19th century, Joseph Fourier showed that we can reproduce, or *synthesize*, any waveform by combining enough sine waves of different amplitudes and frequencies. These sine waves are called *partials* or *harmonics*. The lowest-frequency harmonic is called the first harmonic or *fundamental*. The process of combining harmonics to create new waveforms is called *additive synthesis*.

Given a complex wave, we can synthesize it roughly by combining a small number of har-

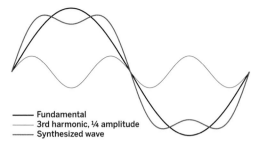

———— Fundamental
———— 3rd harmonic, ¼ amplitude
———— Synthesized wave

J

Adding the third harmonic creates a waveform that has a distinctly square look and sound, though still very rounded.

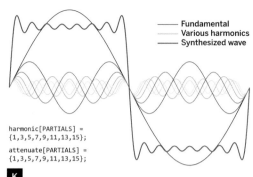

———— Fundamental
———— Various harmonics
———— Synthesized wave

```
harmonic[PARTIALS] =
{1,3,5,7,9,11,13,15};
attenuate[PARTIALS] =
{1,3,5,7,9,11,13,15};
```

K

Adding the first 8 odd harmonics gives a fairly good approximation of a square wave. Individual sine waves appear as little ripples. Combining more partials would reduce the size of these ripples.

monics. The more harmonics we include, the more accurate our synthesis.

Professional additive synthesizers can combine over 100 harmonics this way, and adjust their amplitudes in real time to create dramatic timbre changes. This is beyond the power of Arduino, but we can still do enough to load our wave table with interesting sounds.

Consider the loop in **Listing 1** that calculates a sine wave. Call that the fundamental. To add, say, the third harmonic at ¼ amplitude (**Figure J**), we add a new step:

```
for (int i=0; i<LENGTH; i++) { //
Step across table
  float v = (AMP*sin((PI2/LENGTH)*i));
// Fundamental
  v += (AMP/4*sin((PI2/
LENGTH)*(i*3))); // New step
  wave[i]=int(v+OFFSET);    // Store
as integer
  }
```

In this new step, we multiply the loop counter by 3 to generate the third harmonic, and divide by an "attenuation" factor of 4 to reduce its amplitude.

Listing 3 (available at makezine.com/35) contains a general version of this function. It includes 2 arrays listing the harmonics we want to combine (including 1, the fundamental) and their attenuation factors.

To change the timbre of the sound loaded

into the lookup table, we just alter the values in the 2 arrays. A 0 attenuation means the corresponding harmonic is ignored. The arrays in **Listing 3**, as written, produce a fairly good square wave (**Figure K**). Experiment with the arrays and see what sounds result.

Morphing Waveforms

Professional synthesizers contain circuits or programs to "filter" sound for special effects. For instance, most have a low-pass filter (LPF) that gives a certain "waa" to the start and "yeow" to the end of sounds. Basically, an LPF gradually filters out the higher partials. Computationally, true filtering is too much for Arduino, but there are things we can do to the sound, while it's playing, to give similar effects.

Listing 4 (available at makezine.com/35) includes a function that compares each value in the wave table to a corresponding value in a second "filter" table. When the values differ, the function nudges the wave value toward the filter value, slowly "morphing" the sound as it plays.

Using a sine wave as the "filter" approximates true low-pass filtering. The harmonics are gradually removed, adding an "oww" to the end. If we morph the other way — by loading the wave table with a sine wave and the "filter" table with a more complex wave — we add a "waa" quality to the start. You can load the 2 tables with any waves you like.

Creating Notes

What if we want to make a sound fade away, like a real instrument that's been plucked, strummed, or struck?

Listing 5 (available at makezine.com/35) contains a function that "decays" the sound to silence by steadily nudging the wave table values back toward a flat line. It steps across the wave table, checking each value — if it's more than `127`, it's decremented, and if less, incremented. The rate of decay is governed by the `delay()` function, called at the end of each sweep across the table.

Once the wave is "squashed," running the ISR just ties up the CPU without making sound; the `cli()` function clears the interrupt flag set in setup by `sei()`, switching it off.

Using Program Memory

The Arduino's Atmel processor is based on the "Harvard" architecture, which separates program memory from variable memory, which in turn is split into volatile and nonvolatile areas. The Nano only has 2KB of variable space, but a (relatively) whopping 30KB, or so, of usable program space.

It is possible to store wave data in this space, greatly expanding our repetoire of playable sounds. Data stored in program space is read only, but we can store a lot of it and load it into RAM to manipulate during playback.

Listing 6 (available at makezine.com/35) demonstrates this technique, loading a sine wave from an array stored in program space into the wave table. We must include the *pgmspace.h* library at the top of the sketch and use the keyword PROGMEM in our array declaration:

```
prog_char ref[256] PROGMEM =
{128,131,134,…};
```

`Prog_char` is defined in *pgmspace.h* and is the same as the familiar "byte" data type.

If we try to access the `ref[]` array normally, the program will look in variable space. We must use the built-in function `pgm_read_byte` instead. It takes as an argument the address

Generating Wave Table Data

To read wave table data from program memory, you have to hard-code it into your sketch and can't generate it during runtime. So where does it come from? One method is to generate wave table data in a spreadsheet and paste it into your sketch. We've created a spreadsheet that will allow you to generate wave tables using additive synthesis, to see the shape of the resulting waves, and to copy out raw wave table data to insert into your sketch.

Download it from makezine.com/35.

of the array you want to access, plus an offset pointing to individual array entries.

If you want to store more than one waveform this way, you can access the array in `pgm_read_byte` like a normal two-dimensional array. If the array has, say, dimensions of `[10]` `[256]`, you'd use `pgm_read_byte(&ref[4]` `[i])` in the loop to access waveform 4. Don't forget the `&` sign before the name of the array!

Going Further

Audio feedback is an important way of indicating conditions inside a running program, such as errors, key presses, and sensor events.

Sounds produced by your Arduino can be recorded into and manipulated by a software sampler package and used in music projects

Morphing between stored waveforms, either in sequence, randomly, or under the influence of performance parameters, could be useful in interactive art installations.

If we upgrade to the Arduino Due, things get really exciting. At 84MHz, the Due is more than 5 times faster than the Nano and can handle many more and higher-frequency partials in fast PWM mode. In theory, the Due could even calculate partials in real time, creating a true additive synthesis engine. ◪

Jon Thompson is a UK-based freelance technology writer and managing director of Subversive Circuits Limited. Among other things, he spends his time making strange props for magicians.

EZ-Make Oven

A crafty oven in a paint can.

Written and photographed by **Bob Knetzger**

The poor incandescent light bulb – the very symbol of having a bright idea – is endangered. It's being phased out all around the world: Brazil and Venezuela (since 2005), European Union (2009), and soon in the United States (2014). Granted, efficient CFL and LED lights produce more light with far less energy, but they're just not as much fun.

The light bulb in the EZ-Make Oven creates enough heat to shrink plastic, harden polymer clay, or as in this DIY project, make wiggly creations by "cooking" plastisol. Have your own "aha!" moment and build this light bulb-powered craft oven, while you can!

MATERIALS

» **Paint can, 1gal, empty and clean** with lid and handle. Buy one from the hardware store.
» **Reflector spotlight bulb, incandescent, 75W** Get a spotlight, not a floodlight. Bulb heat varies; I was successful with 65W and 70W bulbs, but you might need higher wattage, so experiment.
» **Lamp bulb socket, ceramic** Look for one with a flanged bottom with mounting holes.
» **Lamp cord, 4'**
» **Lamp cord switch, in-line**
» **Power plug, 110V–120V**
» **Plastisol liquid vinyl** such as Plasti-Goop, Patti-Goop, or #16 Plastisol from iasco-tesco.com
» **Aluminum loaf pan, small**
» **Aluminum tooling foil, 0.005" thick**
» **Sculpey polymer clay**
» **Hardboard, scrap**
» **Grommet, rubber, ¼" ID**
» **Bolts and nuts** to fit your socket
» **Wine corks, synthetic** for feet
» **Wood screws or deck screws, small**

TOOLS

» **Drawing compass**
» **Center punch**
» **Drill with step bit**
» **Deburring tool**
» **Screwdriver**
» **Hardwood dowels, small sizes** for making your own foil tools
» **Sculpting tools**
» **Tweezers or tongs** for lifting hot parts

Unlike thermoplastics that can be melted and remolded, molding with thermoset materials is like making a hard-boiled egg: once cooked, its form is permanently set. What starts out as a syrupy liquid cures into a soft, pliant plastic. Baby boomers might remember the Thingmaker toy from the 1960s. Use this new DIY version to create your own custom mold shapes out of easy-to-form sheet aluminum.

Build the Oven

The oven design is super simple: a light bulb in a can. Holes in the bottom serve as cool air intakes and for mounting the insulated feet and bulb socket. A grid of holes in the lid allows the hot air to rise up around the molds. An inverted loaf pan serves as a cover to hold in the heat.

The oven reaches about 300°F with a 75W spotlight. That's just enough to fuse polymer clay and cure plastisol.

WARNING: The sides of the oven can get hot, so be careful and always use the cool wire handle to lift or move the oven.

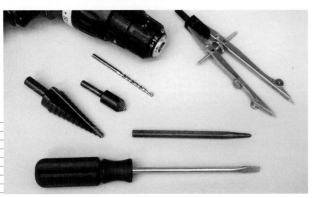

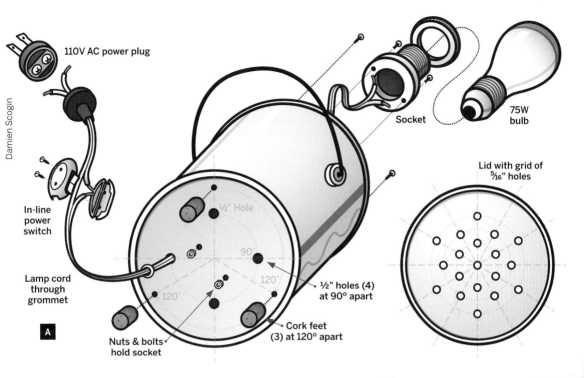

110V AC power plug

Socket

75W bulb

Lid with grid of 5⁄16" holes

In-line power switch

Lamp cord through grommet

Nuts & bolts hold socket

½" Hole

90°

120°

120°

½" holes (4) at 90° apart

Cork feet (3) at 120° apart

A

Damien Scogin

1. LAY OUT THE HOLES

Lay out the holes in the bottom of the can, following the assembly diagram (**Figure A**).

Set the compass to the radius of the can, then use it to mark off distances around the rim to locate 3 equidistant holes for the feet. At each location, mark a spot ½" inside the rim and center-punch each spot.

Place the lamp socket in the center and mark the locations of the mounting holes. Center-punch. (If your socket has a single screw mount in the center, mark and punch a center hole instead.)

Lastly, set the compass for half the can's radius, scribe a circle and mark off 4 equidistant ½" holes in the bottom, and center punch (**Figure B**).

For the holes in the lid, find the center and then use the compass to scribe a circle with a 5⁄8" radius. Use the compass to mark off 6 holes and center punch. Repeat with 1" and 1½" radius circles. Exact spacing isn't critical but you'll want a regular distribution of holes for even heat flow in the center of the lid (**Figure C**).

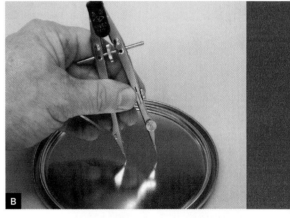

B

C

TIP. A step drill bit works well for drilling in thin metals, where a twist bit tends to grab and pull up the metal. It's easiest to use the step bit with a drill press: control the diameter of the hole by the depth of the feed.

2. DRILL THE HOLES

In the bottom of the can, drill three ⅛" holes for the feet, four ½" holes for venting, and holes to fit the small bolts for mounting your lamp socket (**Figures D and E**).

Drill the grid of holes in the lid (**Figure F**). Deburr all holes as needed.

3. MOUNT THE HARDWARE

Cut 2 synthetic corks in half and drill a ⅛" pilot hole through 3 of the halves. Attach the 3 half-cork feet, using small screws from inside the can (**Figure G**).

Insert the grommet in one of the four ½" holes and thread the lamp cord through it (**Figure H**). Tie a knot in the cord 6" from the end, then split and strip the wires. Connect the wires to the socket. Bolt the lamp socket to the inside bottom of the can using the bolts and nuts (**Figures I and J**).

Install the in-line switch on the cord, 12" from the can, following the directions from your switch (**Figure K**).

Finally, strip the other end and attach the wires to the AC plug (**Figure L**). For a polarized 2-prong plug, the narrow (hot) blade goes to the center (hot) contact on the socket. For a 3-prong grounded plug, the bare or green wire goes to the green ground contact on the socket, if any.

Screw in the bulb and test (**Figure M**).

4. MAKE THE OVEN LID

Invert an aluminum-foil loaf pan and attach a synthetic cork to the outside. Drill a ⅛" pilot hole in the cork and fasten it to the pan with a screw (see **Figure O**, for reference, page 90).

5. MAKE THE FORM

The form is a hard "positive" pattern over which you'll stretch and shape the aluminum sheet to make a "negative" mold. You'll make your form by sculpting soft polymer clay, then firing it.

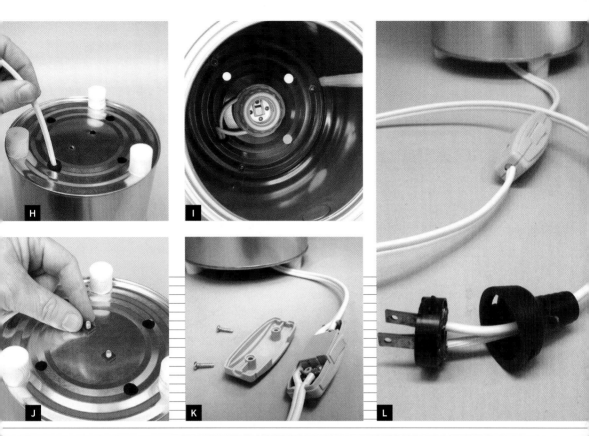

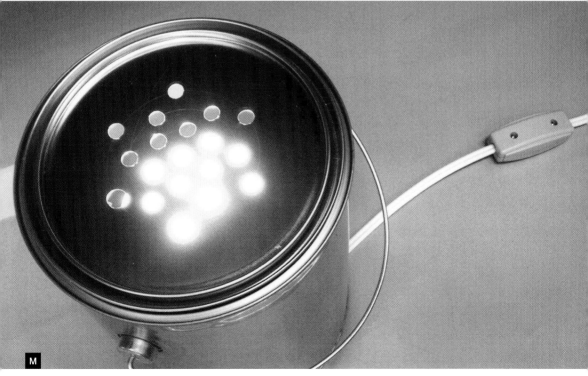

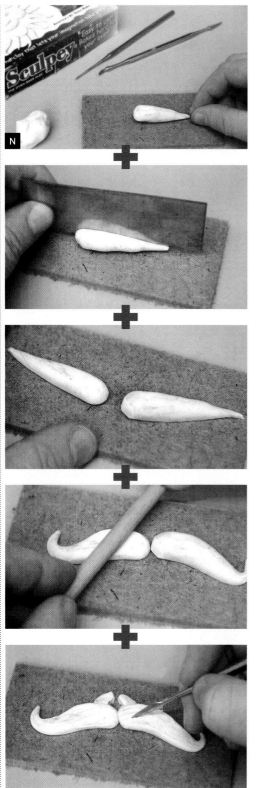

Cut a piece of hardboard on which to make your form. Knead a small quantity of polymer clay until pliant and then sculpt your shape. Keep in mind that you'll be stretching and deforming the aluminum sheet over your form, so keep your shape rather shallow (the aluminum will only stretch so far) and allow plenty of draft (angled sides without undercuts).

Here I'm making a mold for casting rubber moustaches (**Figure N**). Roll the poly clay into a teardrop, then place it on the hardboard. Cut it in half lengthwise with a thin piece of metal or a knife, shape the curl, and add 2 bits to make nose pinchers. Lastly, sculpt in some hair detail —done!

You'll be amazed at the small details that the foil will pick up (like the penny shown in **Figure O**), so really take some time to make small features and textures in your poly clay form.

When your poly clay form is finished, bake it in a 275°F oven for about 20 minutes or until hard. Probe gently with a toothpick; when the clay springs back, it's done.

You can also fire your poly clay using the EZ-Make Oven you just built. It gets hot enough, but it will take more time than using a big kitchen oven. Place the lid on top of the form to keep in the heat (**Figure P**).

When done, set your firm form aside to cool.

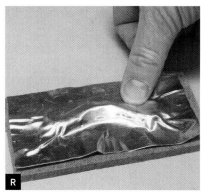

6. TOOL THE MOLD

The material used for the mold must have 3 properties: a low specific heat (to efficiently conduct the light bulb's limited heat to the plastisol), ductility (to be easily formed), and compatibility with the plastisol's chemistry. Steel is too hard, and copper is not chemically compatible, but aluminum has all 3 attributes. Get some 0.005"-thick "tooling" aluminum from an art supply store. It's specially made to be soft and is easily worked with wooden tools. You can also use pieces cut from a disposable aluminum pan. It's a little stiffer and harder to form but will work fine.

Cut a sheet of foil somewhat larger than your form (**Figure Q**). Lay it over the form and press it around the form with your fingers (**Figure R**). A few wrinkles and folds are inevitable, but you'll work them smooth later.

Make some tools from wooden dowels of various diameters. Leave one end flat and make radiused tips on the other end, in various sizes from gently rounded to pencil-point sharp (**Figure S**).

Use the wooden tools very carefully to press the foil closer to the form (**Figure T**). Don't try to stretch the foil all at once; be careful not to tear or poke a hole. Instead, work slowly from the sides or top toward the center of a valley or crease. Start with the roundest-tipped tools first. Do a little at a time. Slowly approximate the shape, stretching the foil as you go. Easy does it! This may take some practice until you get the hang of it.

Once you've defined the basic shape,

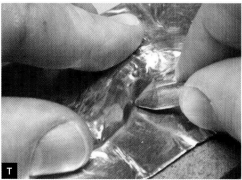

TIP: Often, as you get to the deepest part of the form, where it meets the hardboard, the foil will be very thin and tear easily. Don't despair; your mold may still be quite usable, because you won't fill the mold up to the very top. Keep working!

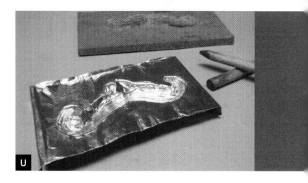

go back with a finer-pointed wood dowel and work the small details. Firm but gentle pressure gives the best result.

Use the flat end or the side of a dowel to smooth out any wrinkles and make a flat surface surrounding your mold shape. Finally, fold down 2 sides to make "legs" so the inverted mold stands level (**Figure U**).

7. CAST YOUR FIRST PART

A good source for bulk plastisol and pigments is Industrial Arts Supply Company (iasco-tesco.com). Get formulation #16 for nicely soft and rubbery molded parts. You can also get ready-to-use precolored "goop" in squeeze bottles from Patti-Goop online at eBay or Amazon. I've even used old bottles of Mattel's original Thingmaker Goop — it still works after 50 years!

Place the mold on the oven and preheat. Fill a squeeze bottle with plastisol and pigment and mix thoroughly. Fill the mold by squeezing in plastisol slowly and letting it fill from bottom to top to avoid trapping any bubbles (**Figure V**).

You can even "paint" your part by using different colors: Let one color set before adding another, or pour together to get a swirl effect.

Place the lid over the mold to help trap the heat (**Figure W**). Thin shapes cure quickly; deep shapes take more time. Probe with a toothpick: when the plastisol is firm it's done. Carefully remove the mold with tongs or tweezers (it's hot!) and let cool.

When cool, use a toothpick to carefully pick at an edge, then lift and peel your part out of the mold (**Figure X**). If you're careful you can cast many parts from the same mold.

8. GO NUTS

There's no limit to the customized or personalized shapes you can create. My daughter Laura molded this cute beetle character based on her own 'zine-comic *Bug Boys*. She added some flexible wires during the casting to give Stag-B bendable, poseable arms and legs (**Figure Y**)!

TIP: If you have trouble getting the oven hot enough, try draping a folded towel over the loaf-pan lid to trap heat, but be safe and keep an eye on it! Need more heat? Try a slightly hotter bulb.

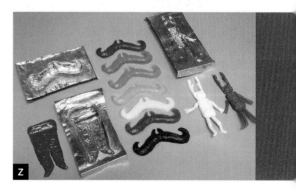

Here's a rainbow of hipster 'staches, Bug Boys, and a tortured tongue (**Figure Z**)! ◪

Bob Knetzger (neotoybob@yahoo.com) is an inventor/designer with 30 years of experience making fun stuff.

1.2.3. Coffee Shop Construction Toy

By **Ian Gonsher** Illustrations by **Julie West**

COFFEE SHOPS ARE GREAT PLACES TO EXPLORE YOUR CREATIVITY and come up with new ideas. A little caffeine, some time to play, and the materials at hand can go a long way. With a bit of imagination you can use this construction method to build many other kinds of forms, and this basic component can be combined with other elements to create larger and more impressive pieces. The baristas don't tend to mind, especially if you tip generously.

YOU WILL NEED: Coffee stirring sticks (12) » Straws (3) » Scissors optional

1. Collect and cut.

After you've customized your caffeinated beverage of choice at the condiment bar, pick up a few extra items: 12 coffee stirrers and 3 straws. Cut the straws in half, creating 6 connector pieces (if you don't have scissors, you can do this with 6 full-length straws as well).

TIP: Soak the stirring sticks in water if you have trouble getting them to bend.

2. Build an element.

Build a basic triangular element with 3 coffee stirrers and 3 straw pieces. Push 2 coffee stirrers, pressed flat against each other, into the end of each of the straws to connect 3 coffee stirrers together.

3. Combine the elements.

Build and attach 3 more triangular elements together until you get a larger triangle. Fold the structure upward, and connect it with the final 3 stirrers. This should give you a cube-like structure. ◢

Ian Gonsher is an artist, designer, and educator on the faculty in the School of Engineering at Brown University. His teaching and research focus on creative practices across disciplines and the development of pedagogical strategies for teaching creative thinking. iangonsher.com

GLASS BEAD
Projection
Screen

⚞ **TIME: 8 HOURS OVER 3 DAYS** ⚞ **COST: $50–$100**

Need a bright surface for your projector? Get high gain at low cost using house paint and sandblasting beads.

Written and photographed by **Sean Michael Ragan**

➕ Titanium dioxide is the most common white pigment in paint, sunscreen, and even food products. It's cheap, safe, and almost unsurpassed in whiteness. It's also the baseline for calculating an optical property called *screen gain*, which is the amount of light reflected from a projection surface divided by the amount of light reflected from a titanium dioxide reference surface. Since titanium dioxide is the pigment used in most white paint, a smooth wall painted flat white has a screen gain very close to 1.

But you can do better. This method applies a high-gain optical projection surface using common, cheap materials – flat white latex paint and glass sandblasting beads. I started out trying to directly mix them (which doesn't work) and happened on this "sprinkling" method by accident. It gives a much brighter screen surface than paint alone.

MATERIALS

» **Hardboard, ¼" panel** sized to fit your projected image; I used 4'×8'. Hardware stores will often cut large panels to size.
» **Lumber, 1×2 (nominal), 8' lengths (4)** or alternate quantities to suit your screen size
» **Wood screws, flat-head, #6×¾" (30)**
» **Wood putty**
» **Paint, flat white interior latex, about 2qts** Plan on 2 fluid ounces per square foot of screen.
» **Glass bead blasting media, 80 grit (25lbs)** such as Harbor Freight item #46426. This process consumes about ¼lb beads per square foot screen area, but considerable excess is needed to ensure complete coverage.
» **Drywall anchors, self-drilling, 40lb capacity, with screws**

TOOLS

» **Measuring tape**
» **Miter box and saw**
» **Hot glue gun**
» **Drill**
» **Drill bit, ⅛" twist**
» **Countersink**
» **Screwdrivers, slotted and Phillips-head**
» **Putty knife**
» **Dropcloth, plastic** 9'×12' is a good size for a 4'×8' screen.
» **Tape** for securing dropcloth
» **Paintbrushes, 1" foam, disposable (2)**
» **Paint roller**
» **Tub or tray, watertight** large enough to allow scooping with dustpan
» **Dustpan** One without a lip is best.
» **Hand broom or brush, medium soft, plastic bristles**

1. Determine your screen size.

Set up the projector as you will use it. Turn it on. Measure the height and width of the image. Plan the size of your screen accordingly. In my case, a single 4'×8' panel made a convenient size.

2. Build the screen.

Cut 1×2 frame members with a miter box and saw. Here's the cut list for my 4'×8' screen.

> » 2 sides 96" long on outside edge, mitered ends
> » 2 sides 48" long on outside edge, mitered ends
> » 2 braces 45" long, square ends

Tack the 1×2s in place on the hardboard with hot glue, then secure with ¾" wood screws every 10" or so. Install the screws from the front side of the unfinished screen, countersink them, and fill the depressions with wood putty.

3. **Lay out the dropcloth.**

A plastic painter's dropcloth will protect your floor, but it's also useful for collecting loose glass beads after you apply the screen surface, so use a fresh one without holes. Spread it in a clean area with a smooth floor, and tape the edges down. Set your screen down in the middle.

4. **Apply the basecoat.**

First paint the edges of your screen with a brush, then apply a smooth, even coat of paint to the surface with a roller. It's easiest to just pour the paint directly from the can onto the screen — Jackson Pollock-style — then smooth it with the roller. Let the basecoat dry 24 hours.

5. **Ready the glass beads.**

Pour out your supply of beads into the tub. A cheap plastic dustpan makes a convenient applicator for sprinkling. With a bit of practice, it's easy to get an even sheet of falling beads.

6. **Apply the topcoat.**

After 24 hours, paint the screen edges again, then pour about ½qt of paint onto the screen, distributing it more-or-less evenly. Roller the paint smooth. You want a quick, even, heavy coat.

7. **Sprinkle the glass beads.**

While the topcoat is still wet, sprinkle beads generously over the entire surface. You'll recover any excess later, so go ahead and apply them all, being careful not to miss any spots. Once the paint is dry, the surface is hard to touch up.

8. **Brush off and recover excess beads.**

Remove the excess beads using a soft brush.

Give the screen back a few thumps to dislodge any

remaining loose beads, then stand it upright for a final brushing. Peel off any flash around the edges with your fingers.

Gather up the tarp from the edges into a "sack," and lift it into your tub. Release one edge and slowly work the tarp out from underneath the mass of beads. I recovered 16 of the 25lbs of beads I applied.

9. Install wall hardware.

I put in a row of 4 self-drilling wall anchors behind the top edge of the screen, and drove in their screws, leaving about ½" sticking out from each. Then I just hung my screen on the screw heads. It's easy to adjust horizontally, but not as secure as I'd like. A French cleat would probably be the best solution.

10. Hang the screen.

Put on clean gloves before handling the finished screen to avoid getting oil on it. Lift it into position and hang it in place.

11. Use it!

Turn on your projector and refocus and adjust the image as necessary.

Conclusion

The final surface — glass beads embedded in latex house paint — is surprisingly tough. I was concerned that flexing the screen would cause the beads to flake off, but the latex paint is still flexible after 2 years. I almost think you could apply it to a thin surface that actually rolls up.

Another pleasant surprise was that, at viewing distances, the surface treatment is remarkably tolerant of small imperfections, and does not require a very smooth texture. I believe it could even be applied directly to a textured wall, thus eliminating the need for a separate screen altogether. ◪

CAUTION: Depending on the size and weight of your screen, you may want to get help lifting it. Be careful!

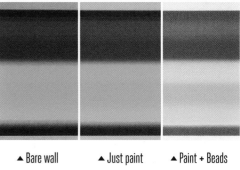

▲ Bare wall   ▲ Just paint   ▲ Paint + Beads

Sean Michael Ragan is technical editor of MAKE magazine. His work has appeared in *ReadyMade*, *c't – Magazin für Computertechnik*, and *The Wall Street Journal*.

Raygun
Vector Weapon

PEW! PEW!

By **Symetricolour**

⚡ **TIME: 2–4 HOURS** ⚡ **COST: $15–$20**

Build a mini analog sound-effects circuit.

Gregory Hayes

The Vector Weapon is a small, lo-fi, electronic noisemaker circuit that makes raygun-style sounds, from classic *pew! pew!* noises to evolving death-ray-type tones. It uses common parts and tools, is easy to build, and has plenty of modding potential. It has a built-in amp chip and speaker so there's no need for external amplification. To hear it in action, check out the video at makezine.com/35.

There are 2 options for construction: a basic version that uses PCB-mounted components, and an acrylic-housed raygun version that uses full-size panel-mount potentiometers and switches – and you can also build it from a kit!

MATERIALS

» **Raygun Vector Weapon Kit**
item #MSVWP from the Maker Shed, makershed.com. Includes laser-cut acrylic raygun case and all parts except 9V battery.

—OR—

» **Stripboard, 24 traces wide × 20 holes high**
» **Schmitt trigger IC chip, CMOS, type 40106 hex inverter**
» **Audio power amp IC, LM386**
» **Chip sockets: 14-pin (1), 8-pin (1)**
» **Battery, 9V**
» **Snap for 9V battery**
» **Speaker, 4Ω** DigiKey #102-1560-ND
» **NPN transistors, 2N3904 (2)**
» **Photoresistor, 5mm** aka light-dependent resistor (LDR) or CdS cell
» **LED, 5mm, red**
» **Resistors: 470kΩ (2), 1kΩ (1)**
» **Capacitors, ceramic: 0.1µF (1), 47nF (1)**
» **Capacitors, electrolytic: 220µF (1), 100µF (1), 4.7µF (1), 2.2µF (1)**
» **Wire, solid core, insulated**

FOR THE BASIC VERSION:
» **Battery spring clip, metal, 9V**
» **Switch, momentary pushbutton** DigiKey #450-1646-ND
» **Trimpots: 1MΩ (2), 100kΩ**
» **Battery spring clip, metal, 9V**
» **Wood screws, round head, #4, ¾" (7)**
» **Scrap wood, 5"×3¼"×¾"** or thereabouts

FOR THE RAYGUN VERSION:
» **Potentiometers, 1MΩ (2), log taper** DigiKey P160KN-0QC15A1M
» **Potentiometer, 100kΩ, linear taper** DigiKey P160KN-0QC15B100K
» **Switch, momentary pushbutton** RadioShack #275-646
» **Wire, stranded, insulated** flexible, not stiff
» **Acrylic sheet, ⅛"×18"×12", any color**
» **Machine screws, #4-40: 1¼" long (6), ½" long (5)**
» **Nuts, #4-40 (11)** locknuts preferred
» **Spacers, #4-40 ³⁄₁₆" long (10)**

TOOLS

» **Electrical tape**
» **Side cutters, small**
» **Soldering iron and solder**
» **Drill**
» **Drill bits: ⅛"; and for raygun version only, ¹⁄₁₆", ⁹⁄₃₂"**
» **Sandpaper, fine**
» **Hobby knife**
» **Laser cutter or jigsaw (optional)** for cutting acrylic, if you're making the case

How It Works

The Vector Weapon is basically a tiny lo-fi analog synthesizer.

One of the 6 inverters on the 40106 IC, a 1M pot, and a 2.2µf capacitor form a square wave oscillator (Osc 1) that mostly runs slower than the main audio oscillator (Osc 2), and is thus considered a low-frequency oscillator (LFO). This square wave is buffered using 2 more inverters and passed through a capacitor to shape it into a rapidly decaying downward slope. The 100K pot works as a voltage divider that determines how much of this signal is sent to a transistor controlling the LED.

Another 40106 inverter, the light-dependent resistor (LDR), the second 1M pot, and the 47nF capacitor form the main audio oscillator (Osc 2). The LED from Osc 1 is optically coupled to the LDR of Osc 2 so the pitch of Osc 2 rises and falls with the output of Osc 1. The LED also adds a little of its own decay as it won't go out instantly. The pot connected to Osc 2 sets its lowest frequency, but the LDR is capable of driving it all the way up into the ultrasonic range. The second transistor "listens" to the capacitor charging and discharging and creates a triangle oscillator at the output.

The signal is finally sent through a power amplifier IC (the LM386) to drive an 4Ω speaker.

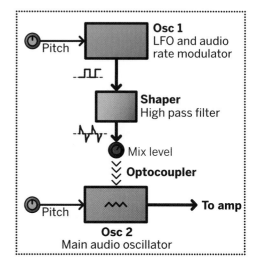

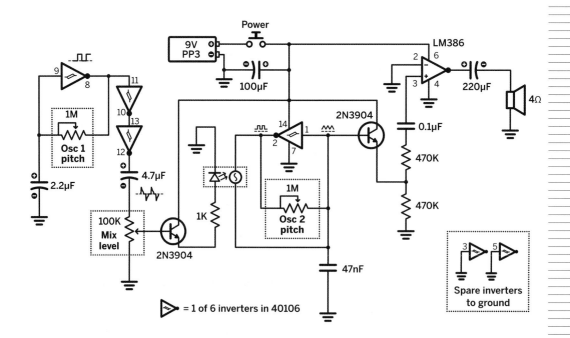

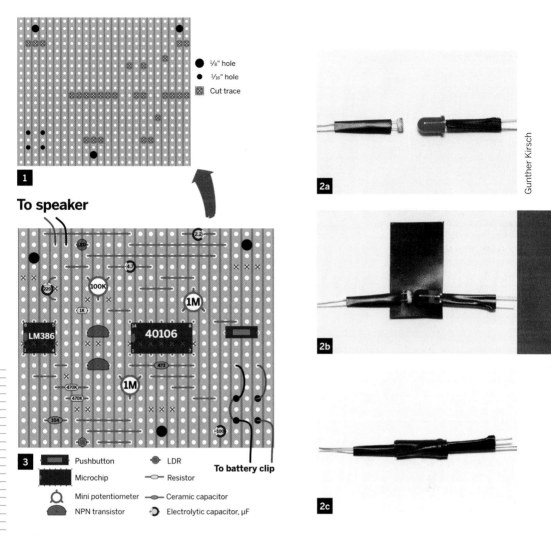

1/8" hole
1/16" hole
Cut trace

To speaker

2a

2b

2c

Pushbutton

Microchip

Mini potentiometer

NPN transistor

LDR

Resistor

Ceramic capacitor

Electrolytic capacitor, μF

To battery clip

1. DRILL OUT THE STRIPBOARD

Figure 1 shows the copper side of the strip-board. The large ⅛" holes are for mounting screws, and the smaller ¹⁄₁₆" holes form a strain relief for the battery connector. Use a ⅛" drill bit or knife to cut the copper traces on the stripboard where shown. Clean off the copper side using fine sandpaper, removing any small flakes of copper and preparing the surface for soldering.

2. BUILD THE OPTOCOUPLER

Wrap the legs of the LDR and LED with electrical tape to insulate them, noting the negative leg of the LED, which is usually the shorter of the two (**Figure 2a**). The LDR is light sensitive, so a few turns of electri-cal tape are needed to make sure it's not affected by ambient light. Connect the two as shown in **Figures 2b and 2c**.

3. SOLDER THE COMPONENTS

If you're mounting the circuit in a case, like the acrylic raygun, you'll need to substitute the trimpots and pushbutton shown in **Figure 3** with 4"-long flexible wire jumpers connected to the panel-mount components (**Figure 5c**, following page). You'll need 6" battery clip leads. Route them through the strain-relief holes before soldering.

Solder in the chip sockets but leave out the chips for now. Be sure the polarized

4

5a

components — the LED, the chips, the electrolytic capacitors, and the transistors — are inserted the right way. The parts are attached to the component side of the board, but **Figure 3** also shows the copper-side traces to make it easier to follow the circuit. It's flipped when compared to **Figure 1**.

The LDR and LED are shown only where they connect to the board, but they actually pass over the amplifier chip, as shown in **Figure 4**.

4. TEST YOUR CIRCUIT

When finished, your board should look like **Figure 4**. Set all the potentiometers to center, attach a battery, and press the button — you should hear something vaguely sci-fi. Adjusting the potentiometers should change the character of the sound dramatically.

If the circuit isn't acting as expected, check it against the stripboard diagram, make sure your solder joints are clean, check that your battery is good, and try again. Note that it's possible to turn the potentiometers up far enough that the circuit stops making audible sound; this is not a fault!

NOTE Make sure you ground yourself before inserting the chips, as static electricity may damage them.

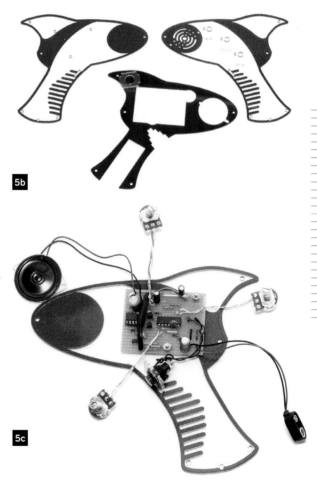

5b

5c

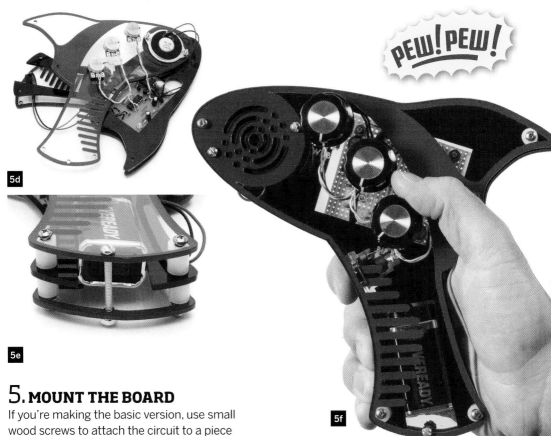

5d

5e

5f

PEW! PEW!

5. MOUNT THE BOARD

If you're making the basic version, use small wood screws to attach the circuit to a piece of wood (**Figure 5a**). Be careful not to tighten too much as you may crack the stripboard. Use two more screws to attach the speaker, another to attach the battery spring clip, and you should be done.

To make the raygun version, download the housing template from makezine.com/35. Use it to cut the 4 case panels out of acrylic, and to locate pilot drill holes. It may be easier to step-drill the large, ½" hole in the small trigger panel before cutting it to size.

Drill six ⅛" holes in the right and left panels where indicated, for the case assembly screws. The middle panel only needs five ⅛" holes, as the sixth screw will pass through the open battery cutout. Drill 3 more ⅛" holes in the right panel for mounting the board, three ⁹/₃₂" (7mm) holes in the left panel for the potentiometers, and 2 more ⅛" holes in the left panel for the speaker mounting screws (**Figure 5b**).

Mount the pushbutton in the trigger panel and solder the leads. Mount the stripboard on the right panel with three ½"-long machine screws and nuts. Align and insert the trigger panel in the right panel slot (**Figure 5c**).

Thread the pot leads through the rectangular opening in the center panel, and the speaker leads through the round opening. Mount the pots in the left panel using their bundled hardware, and the speakers with two ½"-long machine screws and nuts (**Figure 5d**). Snap in the battery.

Guide the 1¼" machine screws through the mounting holes, using spacers between middle and side panels. Add nuts and tighten to close up the case (**Figure 5e**). Attach knobs to the potentiometer shafts and you're ready to conquer the galaxy! ◪

✛ For diagrams, schematics, photos, and video visit makezine.com/35.

Symetricolour designs and builds lo-fi noisemakers, instruments, and kits using electronics and reclaimed materials. To contact Symetricolour, please visit etsy.com/shop/SymetriColour.

I've built a series of Lego phonographs, but none has been as successful as this model. Based on an arm design by Jose Pino (makezine.com/go/pinophone) using a Lego WeDo motor and the Scratch programming environment, it plays loudly enough to be heard across the room with only a 9oz plastic drinking cup serving as the horn. No electronic amplifier is required. It's not exactly high fidelity, and you definitely shouldn't use it with any valuable vinyl, but it's a cool demo and a great learning activity for science, programming, and robotics students. makezine.com/35

Josh Burker works with elementary-age students on constructivist technology projects at an independent school in Connecticut. He has been a Lego maniac since kindergarten.

Meghan Scheck

A Working **Lego** Record Player

Combine WeDo robotics elements, a handful of classic Lego bricks, and a bit of Scratch programming to build a functional phonograph.

Written and photographed by *Josh Burker*

Bookshelf Boombox

CALICO PALACE

Bristow

CROWELL

Hide an amplifier and speaker in plain sight with a project that gives a new meaning to bookshelf speakers.

Written and photographed by
Steve Hoefer

⚡ TIME: A WEEKEND ⚡ COST: $40–$80

✚ Don't judge these books by their covers! They hide an amplifier, a speaker, and a plush compartment for your portable audio player.

Most amplifiers and speakers are flashy — shiny plastic and metal covered with lights, knobs, and indicators — and totally out of place in my reading room. I wanted a design with more warmth and feeling.

Nothing looks as perfect on a bookshelf as books. So the obvious solution was to put a speaker and an amplifier inside a book or three. Putting things in hollowed-out books is a trick as old as bookbinding, but with a little more effort we can even disguise some of the controls in a natural way, such as using a tasseled bookmark as the volume control.

The hidden speaker uses the covers of 3 books to disguise its true nature, with wooden frames providing stability and durability. A muslin cover allows the sound to escape unmuffled while keeping the look of paper. The electronics are based on a commercial amplifier kit that we modify slightly to move some of the components to more useful locations. A padded nook provides a safe place to keep a music player or radio.

MATERIALS

» **Hardbound books, each at least 1¼" thick × 5¼" wide (3)**
» **Plywood, ¼"×12"×24"**
» **Muslin, 12"×36", natural or white**
» **Compact amplifier kit, 5W** Cana Kit #CK550, makezine.com/go/canakit
» **Speaker, 4Ω to 8Ω, 5W, 90mm×50mm** Mouser #665-AS09008PO2WRR, mouser.com
» **Wire, insulated, 22AWG, solid core, multiple colors**
» **Stereo TRS plug, male, ⅛"**
» **DPST switch** RadioShack #275-015
» **Velvet, 24"×24"**
» **Tassel**

For a wall-powered version:
» **Power supply with 2.1mm center positive jack, 9V–16V** Maker Shed #MKSF3, makershed.com
» **Power jack, panel mount, 2.1mm** RadioShack #274-1582

Or, for a battery-powered version:
» **Battery snap connector with leads, 9V**

TOOLS

» **Miter saw/miter box**
» **Craft knife or razor blade**
» **Clothing iron**
» **Ruler**
» **Pencil**
» **Computer**
» **Printer**
» **Paper**
» **Scissors**
» **Combination square**
» **Wood glue**
» **White glue**
» **Multipurpose epoxy, clear or light colored**
» **Sandpaper**
» **Clamps**
» **Drill and drill bits**
» **Coping saw or jigsaw**
» **Soldering iron and solder**
» **Screwdriver, Phillips-head**
» **Wrench, ⁵⁄₁₆"**
» **Pushpins or thumbtacks**

A

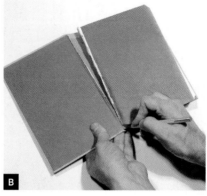

B

TIPS Stores and library sales are good places to buy gently used books. Try to find books that look good without their slipcovers and that you won't mind displaying on your bookshelf.

If you want to keep the pages you can rebind them. See "Olde-School Bookbinding" in MAKE Volume 05.

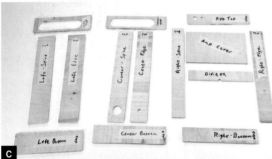

C

D

1. PREPARE THE BOOKS

Find 3 hardbound books that are each at least 1¼" thick and 5½" wide. They work best if they're all the same height; 5¾"×8½" is a common size that will hold everything we need.

Decide on an order in which to shelve the books (**Figure A**). The book on the left will hold half the speaker, the center book will hold the other half of the speaker and the power switch, and the right book will hold the amplifier and storage for a music player.

Remove each cover from its pages by opening the front cover and folding it back (**Figure B**). Use a sharp blade to carefully cut along the joint between the cover and the bound pages.

2. BUILD THE SUPPORT FRAMES

Each book cover is supported by a 4-piece frame. Since your books are probably slightly different thicknesses, measure the width, depth, and height of the pages removed in Step 1 to get the dimensions for the frame for each book. Make paper patterns for each side, being sure to mark each piece with the book

it belongs to (right, center, left), what part of the frame it is (top, bottom, edge, spine), and which direction it goes.

The righthand book has an internal compartment to cover the amplifier board. Cut two more patterns — a "cover" and a "divider" — for the panels that will form this compartment. The cover should be 3" long, and the divider should be as long as the book is thick, minus one plywood thickness. Each should be just wide enough to fit snugly inside the frame, as show in **Figures C and O** (see page 113).

Download the template from makezine. com/35 and use it to add cutouts to the patterns for plugs, speakers, etc.

Cut all the pieces from ¼" plywood following your patterns. Miter the short edges of the frame pieces at 45°, so they fit together at the corners like a picture frame. Be sure to mark each piece, following its pattern, with its position and orientation in the finished frame. Test the pieces for fit, and then use a drill and a jigsaw or coping saw to remove the cutouts (**Figure C**).

Assemble the frames with wood glue and clamps. Use a combination square to keep

E

F

TIP Be careful to keep the glue from getting on the outside of the frame because it will stain.

G

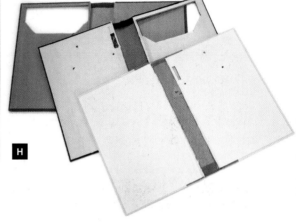

H

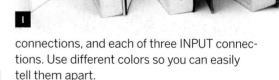

I

the corners true and wait for the glue to dry before removing the clamps (**Figure D**).

For each frame cut a strip of muslin 3" wide and 3' long. Iron out any wrinkles. Glue a single strip of muslin to the outside top, edge, and bottom of the frame. Use white glue, on the inside edge of the frame only, and work slowly around the frame to prevent wrinkles, using pushpins or thumbtacks to hold the fabric in place while the glue dries (**Figure E**).

Use a craft knife to cut an "x" in the cloth over the volume knob cutout. Fold and glue the tabs of cloth inside the hole (**Figure F**). After all the glue has dried use scissors to remove any extra bits of cloth from the inside of the frames. If you're making a wall-powered amplifier, repeat the process with the hole for the DC jack. Your finished support frames should look like **Figure G**.

3. **BUILD THE AMPLIFIER**

Assemble the amplifier kit according to the instructions, but do not solder the stereo jack, DC jack, or speaker output screw terminal.

Solder 10" of 22-gauge wire to the PCB at the DC+ and – connections, both OUT

connections, and each of three INPUT connections. Use different colors so you can easily tell them apart.

4. **CUT THE COVERS**

Use a sharp craft knife to cut holes in the covers following the downloaded template (**Figure H**).

5. **ASSEMBLE**

Attach each frame to the back cover of its book with epoxy (**Figure I**). Again, take care not to let the epoxy get on the outside cloth.

Tip •
Sleeve •
Ring •

Position the power switch, speaker, and amplifier. Route the wires and solder them in place (**Figure K**). The speaker wires are not polarized and can be connected either way to amp OUT.

Wire the stereo TRS plug.
» Solder the INPUT ground wire to the plug sleeve, as seen in **Figure J**.
» Solder the INPUT wire next to C5 to the plug tip.
» Solder the remaining INPUT wire to the plug ring.

Wire the power switch.
For wall power:
» Install the power jack in the matching hole in the frame.
» Connect the center pin of the power jack to one side of the power switch (**Figure L**).
» Connect the other side of the power switch to DC+.
» Connect the outer ring of the power jack to DC−.
For battery power:
» Connect the red wire from a 9V battery connector to one side of the power switch. You might need to add extra wire so it can reach.
» Connect the other side of the power switch to DC+.
» Connect the black wire from the battery connector to DC−.
When all the connections are complete, connect power

Gunther Kirsch (K–M)

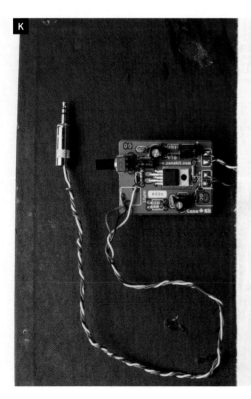

K

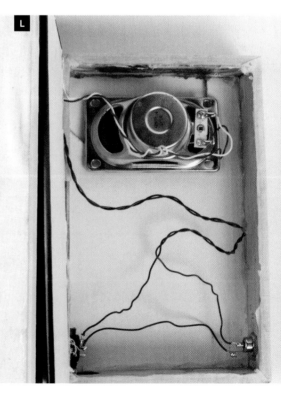

L

and a sound source to be sure everything works correctly.

Use epoxy to join the 3 books cover-to-cover. Epoxy the frame of the lefthand book to its back cover. Use a thin bead of epoxy around the rim of the speaker to fasten it in place in the left and center books (**Figure M**).

Secure the amplifier in the righthand book using the bolt and nut supplied with the amplifier kit (**Figure N**). Use more epoxy to join the center book's front cover to its frame.

Fasten the amp cover and divider pieces in place with wood glue and clamp until dry (**Figure O**).

6. FINISHING TOUCHES

Using your own measurements and the pattern, cut a piece of velvet to fit inside the audio player compartment and to cover the amplifier compartment. Stitch or glue the seams, and use white glue to fasten it in place (**Figure P**).

Place a small amount of epoxy into the edges and top of the volume post, wrap the tassel around the knob, and secure with a knot.

Place the finished boombox on your bookshelf or end table, plug in the AC adapter (or clip on a 9V battery), connect the stereo TRS plug to the headphone jack of your favorite portable music player, and you're ready to listen. ◪

Steve Hoefer is a technological problem solver in San Francisco. He spends much of his time trying to help technology and people understand each other better. grathio.com

TV-Go-Sleep
Universal Timer

⁄ **TIME: A FEW HOURS**   ⁄ **COST: $45–$60**

A remote control to turn off any TV after you've fallen asleep.

Written by
Tom Rodgers

I do a lot of traveling by car, and I usually end up staying in whatever motel is nearest the interstate when I start to feel tired. Often I find that the room has an inexpensive TV that doesn't have a sleep-timer function. So I created the TV-Go-Sleep, a timer that turns off any TV, after whatever delay time I command. You can build it in an afternoon.

The timer is built around an Arduino microcontroller that's been loaded with an open source TV-B-Gone library and a few lines of my own code. A simple pushbutton is used to set the delay time, which is shown on a 7-segment display. When the timer expires, the Arduino uses infrared LEDs to transmit all the TV "off" codes it knows, and the TV shuts down.

MATERIALS

» **Project enclosure, about 5"×2½"×2"** Radio-Shack part #270-1803 or 270-0135, radioshack.com
» **Arduino Uno microcontroller board** Maker Shed #MKSP11, makershed.com. The older Duemilanove also works.
» **PC board or perf board** for prototyping, such as RadioShack #276-158 or 276-149
» **Battery holder, 9V** RadioShack #270-326
» **Battery snap connector, 9V** RadioShack #270-324
» **Switch, SPST rocker** RadioShack #275-693
» **Switch, SPST momentary pushbutton** RadioShack #275-646
» **LED display, 7-segment, common cathode, ½", red** I used Marlin P. Jones & Assoc. part #15119 OP, mpja.com.
» **LEDs, infrared (3)** any size, the brighter the better. I used MPJA #16798 OP.
» **Header pins, male, breakaway** such as SparkFun Electronics #PRT-00116, sparkfun.com. You'll need 16 pins, in blocks of 7, 5, and 4 pins each.
» **Resistors, ¼W: 33Ω (1), 10kΩ (1), 560Ω (8)**
» **Battery, 9V**
» **Hookup wire**
» **Tape, double-sided foam**

TOOLS

» **Wire cutter/stripper**
» **Marker**
» **Saw, small**
» **Soldering iron and solder**
» **Needlenose pliers**
» **Hot glue gun**
» **High-speed rotary tool, or drill and utility knife** I used a Dremel to cut holes in the enclosure, but a drill and utility knife will work as well.
» **Ruler**
» **Solderless breadboard**
» **Masking tape**
» **Computer with Arduino IDE software** free from arduino.cc

1. Install the header pins.

Break off 3 blocks of male header pins and install them in the female headers on the Arduino: a 7-pin block in pins GND through 8, a 5-pin block in pins 6 through 2, and a 4-pin block in pins 5V, GND, GND, and Vin (**Figure A**).

2. Mark and cut the circuit board.

Fit the perf board over the header pins, with its bottom (copper side) facing the Arduino. Sometimes an Arduino's headers aren't lined up perfectly, so you may have to work the board back and forth to get your pins to fit through the holes.

Mark the board where the pins extend through, then trim the board so it's a little smaller than the Arduino beneath it. Leave 1–2 extra rows of holes at each end of the board (**Figure B**).

Test fit, then remove the perf board. Remove your header pins from the Arduino.

3. Build the circuit.

Solder the circuit on the perf board according to the schematic diagram (**Figure C**, following page).

Use the 560Ω resistors for the connections to the LED display. The Arduino's pin 13, used for an all-purpose indicator light, is connected to

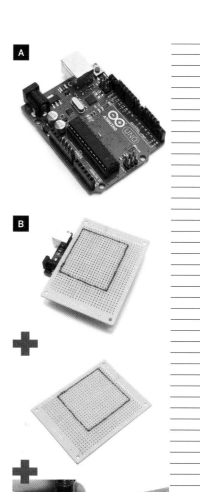

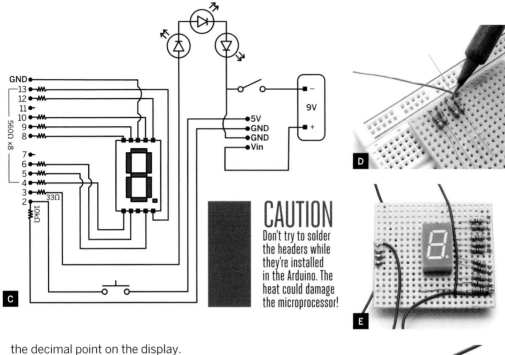

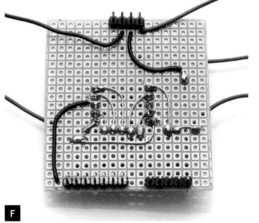

C

D

E

F

Don't try to solder the headers while they're installed in the Arduino. The heat could damage the microprocessor!

the decimal point on the display.

Pin 11 isn't used because it's a PWM pin that shares a timer with pin 3. The TV-B-Gone program uses this timer to create the remote control signals on pin 3, but it accesses the clock directly to increase speed. As a result, an LED attached to pin 11 would flicker when the codes are being sent.

The 33Ω resistor controls the current flowing through the 3 infrared LEDs. I chose its value to maximize brightness and service life. A higher resistance will decrease their range; a lower one will shorten their lifespan.

Pin 2 has 2 connections: to ground via the 10kΩ resistor and to the Arduino's 5-volt reference pin via the pushbutton. When the button is not being pushed, pin 2 is grounded and the input is in the off (or "false") state. When the button is pushed, the pin is connected to the 5V reference potential, so its state is read as on (or "true"). The 10kΩ resistor limits the current flowing between the 5V and GND pins.

Insert the headers from the bottom of the perf board (the copper side) and solder their connections on the top (the component side), as shown in **Figure D**. Use a hot iron and melt just enough solder to get a good

connection — if you use too much you'll end up shorting pins together. It may be helpful to insert the header pins in a breadboard and use bits of wire or masking tape to hold the resistor leads in place while soldering.

Insert the 7-segment display on the top of the board, midway between the 2 header rows (**Figure E**). Make its connections on the bottom of the board, so you can solder the leads to each other and to the copper pad where they meet (**Figure F**).

Cut longer wires for connecting the perf board to the switches and the infrared LEDs.

4. Mount the infrared LEDs.

Drill 3 holes in one end of the enclosure (**Figure G**) and insert the 3 infrared LEDs so they can be wired in series: anode-to-cathode-to-anode-to-cathode. The anode is the longer lead.

Why 3 LEDs? Their lenses focus the IR light into a fairly narrow beam, so adding more LEDs increases the likelihood that one of the beams will reach the TV's IR receiver. Don't make them perfectly parallel; let the lenses point in slightly different directions.

Drip some hot glue on the backs of the LEDs to hold them in place, and then solder them in series.

5. Mount the switches and display.

In the opposite end of the enclosure, make a hole and install your on/off switch (**Figure H**).

In the enclosure's lid, make a rectangular hole for the 7-segment display. The Arduino just barely fits within the width of the RadioShack enclosure, so check the alignment carefully before you start cutting. Leave room for a hole for the pushbutton switch, preferably below the display. Go ahead and cut that hole now, too (**Figure I**).

TIP Masking tape makes it easier to see pencil marks on the lid, and provides a little protection from stray tool marks as well.

Insert the display through the rectangular hole and mount the board under the lid with double-stick tape. Install the pushbutton.

Solder the leads from the perf board to the switches and IR LEDs. Note that the on/off switch interrupts the ground side of the battery. I did it this way to provide an easy grounding point for the infrared LEDs. Hot-glue the battery clip in the bottom of the case (**Figure J**, following page).

6. Install and program the Arduino.

Install the Arduino on the header pins. Be sure to line it up properly!

Use a USB cable to connect the Arduino to a computer (**Figure K**). Download the Universal Sleep Timer Arduino code and the TV-B-Gone for Arduino library (*main.h* and *WORLDcodes.cpp*) from makezine.com/35. Open the files in the Arduino IDE and upload them to the device.

7. **Close it up.**

Disconnect the USB cable, install the 9V battery in the enclosure, and screw the lid in place. I used double-sided tape to attach the instructions to the lid of the box (**Figure L**).

Now Log Some Tube Time

When you turn on the power switch, the decimal point will flash a few times, and then the display will read "9." This corresponds to a sleep delay of approximately 90 minutes. Each time the button is pushed, the sleep delay is shortened by roughly 10 minutes.

If you press and hold the button, the TV-Go-Sleep goes into test mode, and it will begin sending all of its "off" codes.

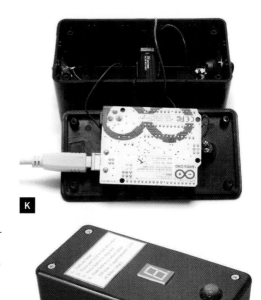

K

L

Place the device where you plan to leave it for the night, and run the test to verify that the TV turns off. If the TV is still on when the test is finished, try realigning the device or moving it closer to the television set.

If the TV goes off and then comes back on, it must be responding to 2 different codes being transmitted from the timer. If that should happen, run the test again, but be ready to press the button when the TV turns off. This will stop the test at that point. Later, when you're asleep and the timer runs down, the device will only send codes until it reaches the point where you stopped the test.

If you can't get the TV to turn off, then its code may not be in the TV-B-Gone library. Check back at righto.com for updated libraries.

Thanks to Donal Morrissey for showing how to put the Arduino into a low-power mode (makezine.com/go/morrissey). And special thanks to Ken Shirriff, who created the Arduino TV-B-Gone code (makezine.com/go/shirriff) that's the backbone of this project. ☒

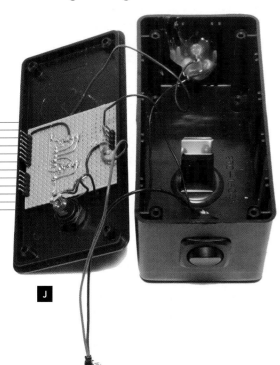

J

Tom Rodgers is a physics teacher and robotics coach in Virginia Beach, Va. He has tinkered with gadgets and gizmos all his life and has subscribed to MAKE since before the first issue was published.

Country **Scientist**

⚡ TIME: 1 HOUR ⚡ COST: $15-20

How to Make and Use Retroreflectors

Written and photographed by **Forrest M. Mims III**

A *retroreflector* is an optical device that returns an oncoming beam of light back to its source. It can be as simple as a tiny glass sphere or a type of prism formed from glass or plastic.

While ordinary flat mirrors also reflect light, the light isn't reflected back to the source but off to the side at the same angle the beam arrived. Only if the light beam is perfectly perpendicular to the surface of a flat mirror does it act like a retroreflector.

Retroreflectors are so much a part of everyday life that typically they don't attract much attention. But they attract plenty of attention while driving at night, when they seem to be almost everywhere. They're incorporated into the taillights of vehicles, safety barriers, traffic signs, and the painted stripes that separate lanes of traffic.

Recently while waiting at a traffic light on a dark night, I noticed seven brightly glowing traffic signs coated with retroreflective paint. These signs were illuminated by the head-lights of my pickup, and they each reflected the oncoming light back toward me.

A nearly full moon was overhead, and it has retroreflective properties, too. That's because the Apollo 11, 14, and 15 astronauts left arrays of precision retroreflectors on the moon. For 40 years, various observatories have pointed powerful laser beams at the moon and detected the light returned by those retrore-flectors to accurately measure the distance between Earth and the moon.

Natural Retroreflectors

Retroreflection was observed long before artificial retroreflectors were invented. That's because the eyes of many animals double as retroreflectors that glow at night when viewed from the same direction as a fire or lamp. When I was a Boy Scout I observed the eyes of alligators in a Florida lake, glowing bright orange in the light of a campfire. Years later I used a flashlight to find caimans in Brazil's remote Cristalino River. Drivers notice the eyes of animals at night glowing brightly in the headlights of their cars.

The retroreflection exhibited by animal eyes is called *eyeshine*. A bright, head-mounted light provides the best way to observe eyesh-ine. Retroreflection from an eye occurs when some of the light focused onto the retina by the lens is reflected back through the lens, where it is refocused into a narrow beam that travels back to the source of light. While this occurs whenever the eye is opened, we notice it only at night when a bright light source held close to our eyes is pointed at the eyes of an animal. The retina alone is not a particularly good reflector, but in many vertebrate animals it is backed by a highly reflective layer of tissue called the *tapetum lucidum*.

The red glowing eyes of people in flash photographs is known as *red eye*. Human eyes lack a reflective tapetum, so human eyeshine is not nearly as bright as that of animals. The best way to eliminate red eye is to move the flash away from the camera's lens. The light will be reflected back toward the flash, and most of it will miss the lens.

Manufactured Retroreflectors

For decades highway signs and painted stripes on roadways have been coated with tiny retroreflectors. The earliest and still the most common are clear glass beads, which can be sprayed or poured over freshly painted road stripes and highway signs. Various kinds of reflective sheets are also used to coat warning barriers and signs. Some employ a layer of glass beads, while others use sheets embedded with tiny plastic corner prisms called *microprisms*.

Microprisms are tiny versions of much larger retroreflectors cut from a corner of a solid cube of glass or silica. These

reflectors are called *corner cubes* or cube corners, and they provide the best performance. The retroreflector arrays on the moon are silica corner cubes. Retroreflectors can also be made by mounting 3 mirrors to form an open corner of a cube.

How to Make Retroreflectors with Glass Beads

Retroreflective glass beads, like those used on highway stripes, are available from various sources. I bought an 8oz bag of standard glass beads for $6 (plus shipping) from colesafety.com.

Any surface that can be painted can be made retroreflective by sprinkling glass beads onto freshly applied paint. Here's how I transformed a plain wood letter "M" from a hobby shop into an attention-attracting object.

1. Place the wood letter on a sheet of paper and apply a thick coat of white enamel.

2. Pour glass beads over the entire painted surface of the letter. Be generous to make sure the entire painted surface is covered.

3. After the paint is totally dry, lift the letter from the paper and gently brush away the loose beads with a finger.

4. Pick up the corners of the paper and pour the unused beads into a container.

5. Place the completed letter in a dark room and shine a flashlight on it from 10' or more away.

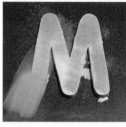

When the light is held near your face, the bead-coated letter will glow bright white.

How to Make an Open Corner Cube Retroreflector

A simple retroreflector can be made from 3 mirrors arranged to form the corner of a cube. You'll need three 2"×2" glass mirrors, double-sided tape, and a standard 2 1/8"×2 1/8"×4 1/8" plastic box, all available from a hobby shop. Use care working with small mirrors, as they have sharp edges and are easily broken; supervise children. Follow these steps:

1. Clean your hands and work surface. Hold the mirrors by their edges and use glass cleaner to clean their surfaces.

2. Place a 1" strip of double-sided adhesive tape on the back of a mirror and insert the mirror, shiny side up, inside the box lid.

3. Place double-sided tape along the lower half of the back of a second mirror and stick it against one of the inner sides of the lid.

4. Stick the third mirror adjacent to the second one so that all 3 mirrors merge to form a corner inside the lid.

5. Look inside the reflective corner of the cube. With one eye closed, your open eye will be directly centered at the apex of the corner. It will stay there even when you move the cube at various angles.

If the sides of the plastic container lid are slightly angled, the reflection from the cube corner will form separate beams that can be seen as 3 spots of light on a white wall behind the light source. When the sides of the box are perfectly square, these separate spots will merge into a single beam.

Going Further

Retroreflectors are ideal for aligning laser intrusion alarms and communication systems.

If you want to make a really big glass bead

CAUTION:
Use suitable laser goggles and avoid eye damage by never looking directly at a laser beam or its reflection.

reflector, see Sean Michael Ragan's projection screen project on page 96. This screen works best when the viewers are seated very close to the projector.

Plastic microprisms are excellent retroreflectors, but they require special equipment to make — until now. It should be possible to make arrays of microprisms using 3D printing with a transparent resin. Since such arrays wouldn't be restricted to flat sheets as with mass-produced microprism arrays, there could be some intriguing applications for 3D arrays of microprisms.

Studies of eyeshine could make interesting science fair projects. For example, populations of spiders and moths that exhibit eyeshine can be surveyed with the help of a head-mounted light on a dark night. Spiders depend on insects for food, so their population is closely related to availability of insects. During a recent major drought where I live, the population of many insects plunged to nearly zero — and so did the spider count. Where there were once dozens of brightly glowing spider eyes each night, only one or two remained during the drought. ◾

Forrest M. Mims III (forrestmims.org), an amateur scientist and Rolex Award winner, was named by *Discover* magazine as one of the "50 Best Brains in Science." His books have sold more than seven million copies.

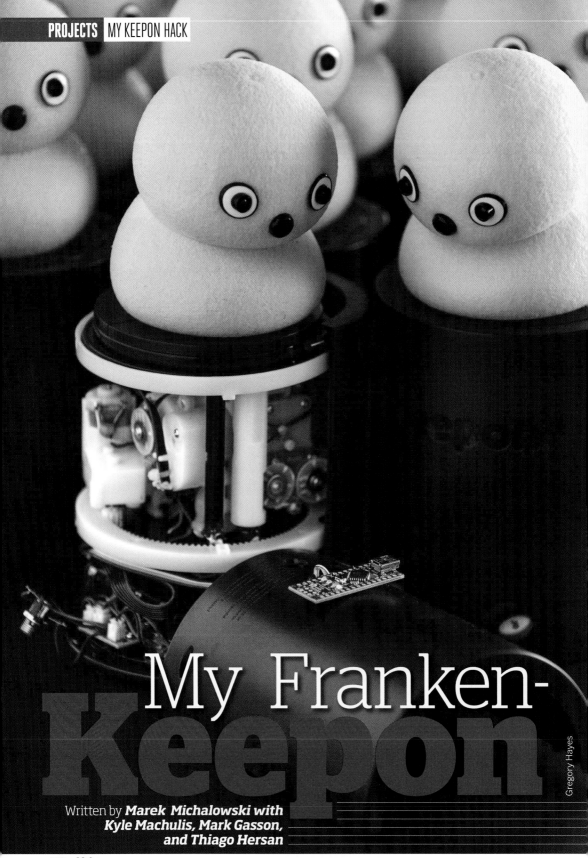

My Franken-
Keepon

Written by **Marek Michalowski** with
Kyle Machulis, Mark Gasson,
and Thiago Hersan

Gregory Hayes

It's alive! Hack your My Keepon toy into a low-cost animatronic puppet with many of the same functions as Keepon Pro.

✗ TIME: UNDER 1 HOUR   ✗ COST: $75

➕ Designed by scientists studying social development, Keepon Pro is a small, "friendly" robot that interacts with children. My Keepon is a low-cost version released by BeatBots and UK-based toy company Wow! Stuff. Unlike the original Keepon, My Keepon lacked the ability to be teleoperated – until now.

These instructions show you how to connect an Arduino to My Keepon's electronics, so you can control it using just about any sensors, devices, and interfaces you can imagine!

The basic principle of this hack is to send commands to the microcontrollers inside My Keepon over the I2C bus. I2C is a two-wire serial interface commonly used for communication between embedded systems and peripherals (wikipedia.org/wiki/I2C). We've provided access to all commands for setting motor speeds and positions, playing sounds, and retrieving information about audio perception, motor EMF, and encoder positions.

Be warned: performing this surgery will void your warranty, so please don't try to take it back to a retailer after doing this. But it's a straightforward procedure, and you'll easily be able to impress other Keepon fans with your new dance choreographies, Kinect mash-ups, and Wiimote-control demos.

In these instructions, we show you how to do this modification in the simplest way, with only a single additional hole drilled into the cylindrical base. You can go further by installing the Arduino Nano board inside the battery compartment, running the USB cable directly into the base, and powering My Keepon with

MATERIALS

- » My Keepon toy
- » Arduino Nano V3.0 without header pins
- » Cable, USB A to Mini-B
- » Cable, 4-wire ribbon, 10"
- » AC adapter for My Keepon, 12V 1.5A, 3.5mm OD, 1.2mm ID, center positive
- » Tape, double-sided foam, 2"

TOOLS

- » Drill or Dremel tool
- » Drill bit, twist, 3mm–4mm
- » Helping hands
- » Hot glue gun
- » Screwdriver, Phillips, small
- » Soldering iron
- » Solder
- » Wire cutter/stripper
- » Safety goggles

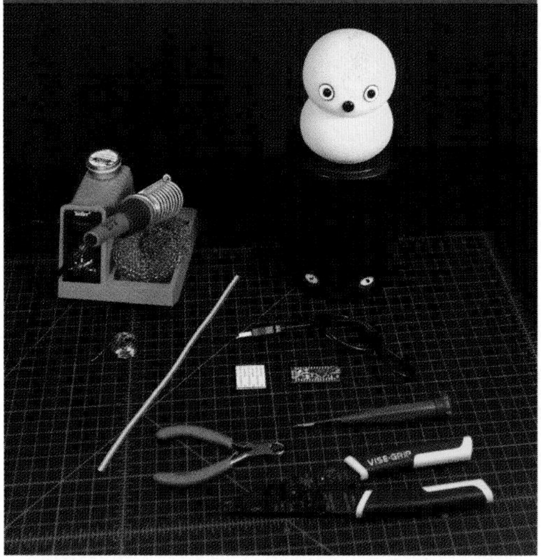

an adapter. But if you want to continue powering the My Keepon with batteries, or if you plan to use Arduino Uno/Mega shields (for wireless communication, additional sensors or actuators, etc.), you'll probably want to "backpack" the components on the base as shown here.

Greg Katz

1a

1. OPEN THE SKULL AND EXPOSE THE BRAIN

1a. Remove the 4 screws securing the back half of the My Keepon cylindrical base. Gently remove the back half of the base, leaving the guts and bottom plate in the front half.

1b. Unplug the ribbon cable running into the control board.

1c. Gently lift the bottom plate (with battery compartment, encoder, and cables) out of the front half. Since there are encoder pins on the control board in contact with the encoder, you may need to lift the robot's mechanical guts a few millimeters up out of the front half; keep the white plastic rings aligned when you do this. Put the bottom plate aside. Remove the clear plastic cover from the control board.

1b

2. ATTACH THE ELECTRODES

Cut 10" or 25cm of 4-wire ribbon cable (or 4 separate 10" lengths of thin/stranded/flexible jumper wire), and strip one end.

Identify the I2C bus pads on the top right corner of the control board, conveniently marked with a friendly silkscreened smiley face. Solder the ribbon cable to the 4 pads, keeping track of which wire you've connected to which pad: V (+ voltage), CL (clock), DA (data), G (ground). You may want to add some hot glue to keep the wire secure.

1c

2

2

3. CLOSE 'ER UP

3a. Widen the hole in the plastic cover to accommodate your new cable. Put the cover back on the control board, running your new ribbon cable through it. Reconnect the original white connector back to the control board.

3b. Place the bottom plate back in the base, lifting the guts slightly if you need to, so that the encoder pins are centered on the encoder.

3c. Drill a hole in the back half of the cylinder, about 1cm above the power port, large enough to accommodate your ribbon cable. Thread about 1" of your ribbon cable through the hole.

3d. Place the back cover on the cylinder, carefully tucking away a fold of your ribbon cable so that it will have room to rotate with the mechanism. Replace the screws securing the back of the cylinder.

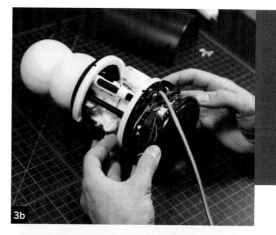

3b

3c

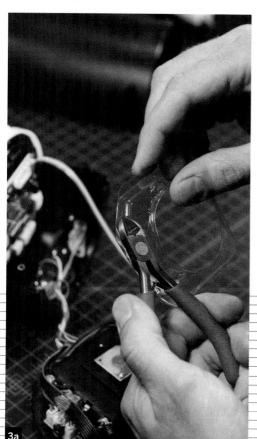

3a

3d

4a

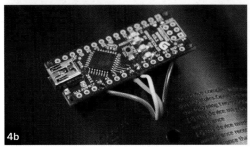

4b

4. ATTACH THE ARDUINO

4a. Strip 2mm–3mm of the ribbon cable protruding from the robot and solder the 4 wires to the appropriate pins on the Arduino. My Keepon's V pin connects to the Arduino's A0 pin; CL connects to the A5, DA connects to A4, and G connects to GND.

4b. Secure the Arduino to the back of the robot however you prefer (e.g. with double-sided foam tape or hot glue).

5. PULL THE STRINGS

With My Keepon powered off, connect the Arduino to your computer with the USB cable. Flash the Arduino with the *MyKeepon.ino* source file at github.com/beatbots/mykeepon. Start the Arduino Serial Monitor (or another application written to talk to the robot) at 115,200 baud. When the Arduino's code begins running, it waits to see voltage on the A0 pin (from My Keepon's V pin).

Now power on My Keepon; the Arduino can now receive serial commands and translate them to the appropriate I2C messages for My Keepon's controller. Happy hacking! ↗

➕ To see these guys in action, check out the video at makezine.com/35.

GOING FURTHER

At our Github page you can find a Max/MSP control patch, a standalone application built from the patch, and an openFrameworks-based application for controlling multiple My Keepons. But you can send the command strings from any application or code of your own devising. Allowable command strings are:

```
SOUND PLAY <0...63>;
SOUND REPEAT <0...63>;
SOUND DELAY;
SOUND STOP;
SPEED [PAN, TILT, PONSIDE] <0...255>;
MOVE PAN <-100...100>;
MOVE TILT <-100...100>;
MOVE SIDE [CYCLE, CENTERFROMLEFT,
   RIGHT, CENTERFROMRIGHT, LEFT];
MOVE PON [UP, HALFDOWN, DOWN, HALFUP];
MOVE STOP;
MODE TEMPO;
MODE SLEEP;
```

The Arduino also periodically sends back data strings from My Keepon's I2C bus (you can change their frequency in the Arduino code):

```
BUTTON [DANCE, TOUCH] [OFF, ON]
BUTTON [HEAD, FRONT, BACK, RIGHT,
   LEFT] [OFF, ON]
MOTOR [PAN, TILT, SIDE, PON] FINISHED
MOTOR [PAN, TILT, SIDE, PON] STALLED
ENCODER TILT [NOREACH, FORWARD,
   BACK, UP]
ENCODER PON [HALFDOWN, UP, DOWN,
   HALFUP]
ENCODER SIDE [CENTER, RIGHT, LEFT]
ENCODER PAN [BACK, RIGHT, LEFT,
   CENTER]
EMF [PAN, TILT, PONSIDE] [-127...127]
POSITION [PAN, TILT, PONSIDE] [VAL]
AUDIO TEMPO [67, 80, 100, 133, 200]
AUDIO MEAN [0...64]
AUDIO RANGE [0...64]
AUDIO ENVELOPE [0...127]
AUDIO BPM [VAL]
```

Dr. Marek Michalowski is a roboticist living and working in San Francisco. He co-founded BeatBots with Dr. Hideki Kozima (designer of Keepon). See more at beatbots.net.

1.2.3. Simple Light-Up Hoodie

By **Craig Couden** Illustrations by **Julie West**

TRON: LEGACY **WAS A DECENT MOVIE, BUT WHAT I WAS REALLY WOWED** by were the luminescent costumes. I've been playing around with light-up clothing ideas ever since, and the simplest idea uses EL tape and some velcro to approximate Jeff Bridges' glowing druid look from the film. It's not remotely screen accurate (for you purists out there), but it is a great way to wow your friends during a night out.

YOU WILL NEED: **Sweatshirt** with large hood
» **EL tape strip, about 40"** Adafruit #415
» **Pocket-sized EL inverter** 4×AAA or 2×AA
» **EL wire extension cord** Adafruit #616
» **Velcro** enough to match the EL tape
» **Batteries, rechargeable** » **Sewing machine or needle and thread** » **Glue** optional

1. Sew velcro to hood.

Cut a length of velcro equal to the length of the EL tape and hand or machine stitch the soft side onto the inside of the hood near the edge. You may want to remove the drawstring first.

2. Attach velcro to EL tape.

Attach the prickly "hook" side of the Velcro to the back of the EL tape. (Mine came with an adhesive backing, but you may have to glue it.)

3. Juice it.

Attach the EL tape to the hood. Connect the EL tape to the extension cable, then the inverter. Turn it on, stick the inverter in your pocket, and show that you, too, fight for the User!

Going Further

To achieve a longer wear time, use 2 shorter lengths of EL tape that meet at the apex of the hood, each with its own 2×AA inverter. ▧

When not obsessing over his new couture EL fashion line, Craig Couden is MAKE's editorial intern. He also writes about board games at dragonstormgames.com.

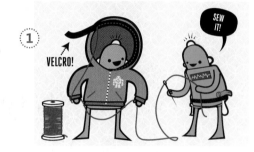

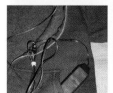

**Don't have an iron?
Use a stamp to
burn your mark!**

Chemical

Written and photographed by **Sean Michael Ragan**

Wood Burning

↗ **TIME: 1 HOUR**   ↗ **COST: $20**

➕ If you want to apply a maker's mark or other repeated pyrograph to wooden goods but can't justify the expense of a custom branding iron, an indistinguishable effect can be achieved by applying a strong solution of ammonium chloride, for instance using a rubber stamp, then cooking it with a heat gun.

When heated, ammonium chloride decomposes into ammonia gas and strong hydrochloric acid ($NH_4Cl \rightarrow NH_3 + HCl$).

Ammonia diffuses away into the air, leaving the acid behind, which burns the wood. The resulting chemical burn is identical to a heat burn in most respects.

This process sounds nastier than it really is, and although prudence dictates erring on the side of caution and working with plenty of ventilation, the process does not produce a noticeable smell either of ammonia or of HCl. The only detectable odor is burning wood.

MATERIALS
- » **Ammonium chloride, 25g, aka "sal ammo-niac"** Get the powdered form. Solid bricks sold for tinning soldering irons can also be broken up in a mortar and pestle.
- » **Water, 2 cups**
- » **Wooden workpiece** to be burned

TOOLS
- » **Felt, 8"×10"**
- » **Heat gun**
- » **Measuring cup, 2c/500mL**
- » **Measuring spoon, 1Tbsp**
- » **Spray bottle, 500mL**
- » **Stamp, rubber** I cobbled one together from foam rubber letters, a scrap of MDF, and carpet tape. Solid rubber stamps work too.
- » **Tray, aluminum or plastic** to fit stamp
- » **Sheet of paper**
- » **Gloves, latex or nitrile**
- » **Safety goggles**

CAUTION: Ammonium chloride fumes can cause respiratory harm or an asthma-like allergy, so avoid breathing dust, vapor, mist, or gas. Wash thoroughly after handling. Store ammonium chloride capped at room temperature and protect it from heat.

A

1. PREPARE A STAMP
For my simple "MAKE" stamp, I used these plastic-backed 1½" foam-rubber letters from a hobby store (**Figure A**).

Apply a strip of carpet tape to a scrap of MDF, plywood, or other flat stock. Remove the tape backing (**Figure B, next page**). Arrange the letters as needed and push each down to fix it in place (**Figure C**).

An ordinary rubber stamp should also work with this process.

B

C

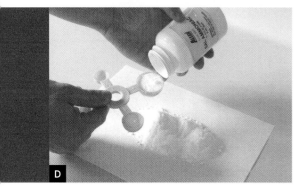

D

CAUTION:

Although these photos show me working with bare hands, use caution and wear latex or nitrile gloves, as well as goggles.

E

NOTE:

If you want more or less than 500mL of "ink," the formula is 1 tablespoon ammonium chloride per 100mL water.

F

2. MEASURE THE POWDER

Fold a sheet of paper in half lengthwise, then measure 5 level tablespoons of ammonium chloride onto the fold (**Figure D**).

Using the crease as a channel as shown, pour the powder into the mouth of the bottle (**Figure E**).

3. ADD WATER AND MIX

Pour 500mL of warm water into the bottle on top of the solid ammonium chloride. Tap water should be fine (**Figure F**).

Put the sprayer, or a matching cap, on the bottle and tighten it down securely. Shake the bottle gently until all the ammonium chloride is dissolved.

4. PREPARE A STAMP PAD

Set out a flat, shallow, aluminum or plastic tray that's big enough to accept your stamp. Line the tray with felt.

Saturate the felt by misting it generously with ammonium chloride "ink" from the spray bottle (**Figure G**).

5. INK AND STAMP

You'll want to practice this process on scrap wood before trying it on anything important.

Put the stamp onto the felt and press down with some force to "ink" it (**Figure H**).

Transfer the stamp to the workpiece, line it up carefully, and press down with about the same force to transfer the ink (**Figure I**).

6. APPLY HEAT

Immediately after applying the stamp, pick up the heat gun and begin applying heat. I used the "high" setting on my heat gun.

Play the heat gun evenly across the surface of the work. Within a minute or two the inked areas will begin to turn yellow, then brown, then brownish-black (**Figure J**).

You're done! For all practical purposes, the resulting chemical burn is indistinguishable from a heat burn. It's waterproof and can be finished or otherwise treated like a conventional pyrograph.

Going Further

The "ink" described here is nothing more than an 80% saturated solution of ammonium chloride. Though it works well enough, there's plenty of room for improvement. Adding egg whites or other thickener, for instance, might improve the handling qualities by reducing the tendency of the ink to diffuse along grain lines.

And there's no reason it has to be applied with a stamp. You might just as well use a stencil, marker, brush, inkjet printer, or some other means.

Other surfaces should also be susceptible to this treatment. It will certainly work on paper and cardboard. Leather, too, perhaps? Metal? Let us know what you find out at blog.makezine.com/projects/chemical-woodburning. ✂

Sean Michael Ragan is technical editor of MAKE magazine. His work has appeared in *ReadyMade*, *c't – Magazin für Computertechnik*, and *The Wall Street Journal*.

Ctesibius and the
Written by **William Gurstelle**
Tantalus Cup

⚡ **TIME: 1 HOUR** ⚡ **COST: $10-$15**

MATERIALS

» **Plastic wine cup, stemmed** I used the Red Cup 8oz Wine Goblet, Amazon #B009DMEHVE. If you use a different cup, make sure it's got a stem wide enough to drill out and accept the siphon.
» **Silicone glue/ sealant**

Option A:
» **Pill bottle, 2½" high**
» **Tubing, copper or glass, ¼" OD, 2½" long**

Option B:
» **Tubing, flexible plastic, ¼" OD, 6" long**
» **Tubing, aluminum, 3/16" OD, 1" long**
» **Tie-wrap, small** aka cable tie or zip tie

Option C:
» **Tubing, soft glass, ¼" (6mm) OD, 6" long** such as chemistry tubing

TOOLS

» **Electric drill with ¼" bit**
» **File**
» **Hacksaw** if you're cutting metal tubing
» **Scissors** if you're cutting plastic tubing
For Option C only:
» **Gloves, heavy and heatproof**
» **Torch, propane or butane**
» **Rod or pipe, metal, ¾" diameter, about 6" long**

Decipher the siphon and build a high-class practical joke.

While Ctesibius may not be as famous as Archimedes, Aristotle, and Pythagoras, he certainly deserves recognition as one of the world's first and best engineers. A prolific inventor living in the third century BC, Ctesibius is often referred to as "the father of pneumatics." He created the first truly accurate water clock, a compressed-air catapult, a water pump, a musical organ that ran on water, and a number of remarkable machines that made use of siphons.

At one time historians credited Ctesibius with inventing the siphon, but we know it's much older. Around 1400 BC, during the reign of Pharaoh Amenhotep II, Egyptian engravers etched pictures on the walls of tombs at Thebes. One of these shows a group of aficionados siphoning wine from several containers into a large punch bowl, presumably to produce a blend of superior flavor and bouquet. The ancient Egyptians probably also used the siphon for purifying water, irrigating crops, and drinking beer.

But it was the classical Greeks, and Ctesibius in particular, who really explored the scientific principles by which siphons work. We know few details about Ctesibius' life. We do know that he was the son of a barber, and he rose from humble beginnings to become the head of the Library of Alexandria, which at the time was the greatest on Earth.

Ctesibius seems to have had a particular fondness for hydraulics and pneumatics, as well as a wry sense of humor. He put the siphon to work in a number of amusing inventions, including mechanical singing birds, a religious statue that could alternately stand and sit while being carried through the streets, and the "engibita," apparently automatons that could seemingly drink water.

Siphons are sort of mysterious devices. Most people, if they have an opinion on them at all, believe that siphons work because of a difference in air pressure between the upper and lower levels of the water reservoirs. But that's incorrect, because a siphon works even in a vacuum. (There's a terrific video at makezine.com/go/siphon that shows this.)

So, if air pressure isn't the reason, then how does a siphon work? There are two factors: the force of gravity and the attraction between molecules in a liquid, which is called cohesion. To understand this, think of a siphon as an inverted U with one leg longer than the other. The weight of the fluid in the longer leg is obviously greater than the weight of the fluid in the shorter leg. As the fluid in the longer leg falls out due to the pull of gravity, the molecules in the shorter leg are acted upon by two forces: gravity and cohesion. In a siphon, molecular attraction wins out over gravity and pulls the fluid up and over the hump.

The Tantalus Cup

The Tantalus Cup combines the workings of a siphon with a droll sense of humor. Doubtlessly Ctesibius would have found it amusing. It's sometimes called a Greedy Cup or Pythagoras Cup despite the fact that Pythagoras certainly did not invent it.

The cup works normally when partially filled. But if the user fills the cup beyond a particular level, a hidden siphon empties the cup. Far cleverer than a run-of-the mill dribble glass, it's a combination science lesson and practical joke in one easy-to-make package. Ancient versions had a small figure of Tantalus inside; alas, wine would never reach his lips.

You can make the siphon 3 ways (**Figure A**). Option A is easier, but Options B and C have a nicer appearance.

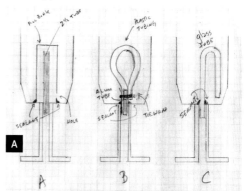

A

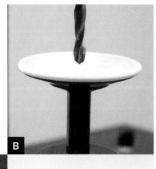

B

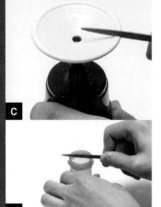

C

D

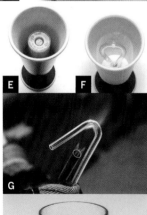

E   F

G

H

1. PREPARE THE CUP

Drill a ¼" hole lengthwise through the stem of the drinking cup (**Figure B**).

Next, file several openings in the bottom lip of the cup, to provide an exit for the liquid when it drains out (**Figure C**).

2. MAKE THE SIPHON

Option A: Pill bottle. As shown in Diagram A, Figure A (previous page), insert the 2½" tube about ½" into the hole you drilled. Fix it in place with the silicone sealant, then seal the joint all around.

File a ¼"×¼" opening in the lip of the bottle (**Figure D**). Apply sealant to the lip (don't plug up the opening you just cut), invert the bottle over the tube, and press it into the bottom of the cup (**Figure E**). Let dry.

Option B: Plastic tube. Following Diagram B, insert the 1"-long aluminum tube halfway into the plastic tube, then insert this assembly into the hole in the cup. Seal the joint with silicone and let dry.

Bend the plastic tube into a loop, taking care not to kink it, until its open end is ¼" above the bottom of the cup. Fix with a cable tie (**Figure F**).

Option C: Glass tube. Buy a 6" length of glass tubing if possible. If not, use the edge of a steel file to notch the tubing, wear heavy gloves to break it to length, then sand or fire-polish the sharp broken ends.

Mount and light a propane or butane torch securely, donning heatproof gloves. Referring to Diagram C, rotate the area of the tube to be bent in the center of the flame until it softens. Remove and quickly bend it around the metal rod to form a 180° U (**Figure G**).

Once the tube is cool, insert it into the hole in the cup and seal the joint with silicone. A polycarbonate wine glass adds some class (**Figure H**).

Using the Tantalus Cup

The legend of the Tantalus Cup is that it was designed to teach moderation. The cup appears normal enough, except for the column or figure inside.

If you pour a moderate amount of wine into the cup, you can drink out of it in the normal fashion. But if you fill it to the brim (actually, to any level over the height of the siphon assembly) the cup will empty itself (**Figure I**). Every last drop will siphon out, and you won't get a drink at all. ◪

William Gurstelle is a contributing editor of MAKE. The new and improved edition of his book *Backyard Ballistics* has just been released.

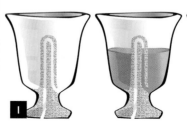

I

My Electro-Chemical
Kid Sub

Written and photographed by
Michael Lewis Wernecke

↗ **TIME: 7 HOURS**   ↗ **COST: $40**

Finally — the baking-powder submarine gets an electric motor!

The 1950s were a golden age for the curious kid. One day, my friend Merle and I fished one of those free plastic submarines out of a cereal box. Up and down it went, in a small container of water, with only a pinch of baking powder to create an air bubble for buoyancy. We loved it.

Fifty years later, I wondered if I could advance the design and inject a little more fun. The answer is an unequivocal yes!

I built the Kid Sub from plastic energy drink bottles, fast-food knives, urethane foam, and various bits around the shop. Like the classic cereal-box toy, it dives and ascends using baking powder to create bubbles for flotation. But this sub's also got forward propulsion from an electric motor, so it can perform powered dives, ascents, and turns, like a real submarine.

The circuit's voltage is so low (1.5V) that immersing the switch and wires doesn't impair the sub or present any danger. What does have to be waterproofed is the motor — it's sealed in petroleum jelly and beeswax.

Dive planes are positioned to control the submarine's angle of ascent or descent, while a rudder controls left-right turning. At the water's surface, the bubble chamber fills with water, decreasing the sub's buoyancy and causing it to sink. The chamber's vent is plugged by a stopper that's forced backward by the sub's forward motion. Inside, contact with water causes Alka-Seltzer to release carbon dioxide (CO_2) gas. The resulting air bubble forces water out of the chamber, increasing buoyancy, causing the sub to rise.

The sub breaks the surface, relieving water pressure on the stopper lever. The air bubble escapes. Water fills the bubble chamber, causing the sub to sink again. Repeat!

+ Build your own Kid Sub: complete tutorial at makezine.com/35.

Michael Lewis Wernecke (ocean_tech04@yahoo.com) holds a master's degree in fine arts from the Otis Art Institute in Los Angeles. He has been a model builder for Disney films, and a senior machine builder for the Whitehead Institute's Human Genome Project. He is an avid reader, nature lover, inventor, and a proud veteran of the United States Army.

DIY
Baseball
Tees

Build a pair of adjustable, durable batting tees for half the price of store-bought.

⟋ **TIME: 1 HOUR** ⟋ **COST: $40**

Written by
Brad Huffman

MATERIALS

» **Floor flanges, galvanized iron, 1¼" (2)** such as Home Depot part #182141
» **Plywood, ½" or ¾", 10"×10" squares (2)** Scrap is fine.
» **Furniture leg tips, rubber, 1½", white (2-pack)** I used Shepherd #89225 from Home Depot.
» **PVC pipe, Schedule 40, 1¼" diameter, 13" lengths (2)**
» **PVC pipe fittings, Schedule 40, 1¼", slip to MIPT male threaded adapter (2)** such as Mueller #436-012HC
» **Disposer hose, 7/8" ID × 1¼" OD, 2' lengths (2)** Pre-cut lengths such as Watts #WPMX2 are nice and straight.
» **Wood screws, ¾" (8)**
» **PVC cement (optional)**

TOOLS

» Saw
» Screwdriver
» Utility razor knife
» Drill and 1⅛" Forstner bit (optional)

➕ Baseball season is here, and my son William has moved up from tee-ball to coach-pitch league. Even though he won't be using the batting tee in games anymore, it's still a great tool to use in hitting practice. When our old tee wore out, I decided to see if I could put together a great tee for less than I could buy one. The answer: Yes!

1. MAKE THE BASE

Center the floor flange on the plywood square and fasten it with wood screws.

2. MAKE THE STEM

Cut the PVC pipe to 13" long. Using a utility razor knife or a drill with a Forstner bit, cut a 1⅛" hole in the end of the rubber furniture leg tip. Work slowly; it's thick stuff. You're looking for a snug friction fit with the 1¼" disposer hose, so be conservative. On the leg tips I bought, there are concentric ridges; one of them fell just within my 1⅛" circle, so I used it as my cutting guide.

Now cram the PVC pipe into the furniture leg tip. This might be the hardest part of the build — it doesn't want to fit, you're really stretching and forcing it. Once you get it started, you can flip it over, put the leg tip on the floor, and press the pipe the rest of the way in.

Push the other end of the pipe into the male threaded adapter. You can use PVC cement if you want, but a strong friction fit should do the trick.

3. MAKE THE TELESCOPING BALL REST

Cut the 1¼" disposer hose to 2' long and stuff it into the hole you made in the chair leg tip. You want a tight fit, so when you adjust the height it stays put without slipping down. If necessary, widen the hole slightly to achieve the right fit.

Use the razor knife to carefully shave a bevel on the inside edge of the hose, so that it will cup the baseball and keep it more stable.

4. ASSEMBLE

Screw the stem's threaded adapter into the flange on the plywood base. You can unscrew it for easy packing and transport.

Conclusion

This project specifies materials for making 2 tees, because I like to do double-tee batting drills that help improve my players' swing technique.

Version 1 of this tee had a 24"-long PVC pipe. I hit off it for a while, and it proved to be relatively stable. As for durability, I was asked whether the PVC needed to be shorter to avoid being broken by an errant bat, so I tested it out. A good whack shattered the top of the tee. The "repair" was pretty easy. I sawed off the broken end and discovered that Version 2 worked just as well with a 13"-long pipe, with less chance of it being struck. It's probably more stable too, due to the lower center of gravity. That's the nice thing about a DIY project — if you do break the PVC pipe,

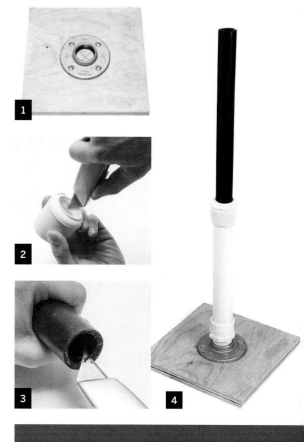

TIP: Make your tee even more stable by slipping your batting weights (donuts, etc.) over the top to rest on the base.

you can easily replace it at negligible cost instead of buying a new tee.

I used V1 and V2 for almost all my hitting sessions last baseball season and for several games, and V2 seems plenty durable with the shortened PVC. The disposer hose has plenty of give so the tee doesn't tip over every time you knock it. ◪

Brad Huffman is a software developer by trade, with a genetic "I can make that" attitude inherited from his father and grandfather. He has amassed a nice collection of tools, parts, and ideas for fun projects, like his Arduino-powered launch controller for MAKE's compressed air rockets; check it out at bradhuffman.com/wordpress.

Gunther Kirsch

Written by
Sean T. McBeth

Making
Pigments

⚡ **TIME: 1 HOUR OVER 2 DAYS**   ⚡ **COST: $10–$20**

Use household
materials and
simple electrolysis
to make metal
oxide powders
for tinting paints.

Illustrated by Nate Van Dyke; Photograph by Gregory Hayes

After reading about ferrofluids (liquids with magnetic properties) in MAKE, I started trying to make my own. I learned how to electrolyze water with iron electrodes to create magnetite – a jet-black iron-oxide – and realized the same process could be used to make simple metal oxide pigments. Instead of just buying supplies at the art store, I wanted to make a painting truly "from scratch."

Electrolysis is the process of using direct current to cause a chemical reaction that would not otherwise occur. It requires an electrolyte, 2 electrodes, and a current source. It has many applications, but the most common is probably electroplating, which is used, for example, to coat a large piece of cheaper metal with a small amount of a more expensive one that is more attractive or a better conductor.

My pigment-making process is dirt simple. The current source is an old cellphone charger with the plug cut off. The electrodes are pieces of common metal hardware, and because they're identical I don't have to pay attention to the current's polarity. The electrolyte is just salt water.

The reaction takes 3 hours and produces a brownish-black mixture of iron oxyhydroxides. It can be repeated with copper electrodes to give it a vibrant orange.

Gunther Kirsch

1

1. Mix the Electrolyte.

Water on its own is a poor conductor, so we add an ionic chemical — in this case sodium chloride — to increase its conductivity. Pour a small amount of salt into the glass dish. Fill about halfway with water and stir to dissolve.

2. Wire the Electrodes.

Cut off the charger plug and strip the wires (**Figure 2a**, following page).

Twist the wires around the screws near the heads (**Figure 2b**).

Position the screws at opposite edges of the dish, keeping the wire above the waterline.

MATERIALS

» **Steel drywall screws (2)** Actually, any 2 pieces of carbon steel or iron should work.
» **Tape**
» **Water**
» **Table salt**
» **Acetone**

TOOLS

» **Wire cutter/stripper**
» **Dish, glass**
» **Cellphone charger rated 5V/700mA**
» **Gloves, acetone-resistant** neoprene rubber
» **Funnel**
» **Paintbrush, small, plastic**
» **Rare earth magnet** the stronger, the better

SAFETY: This process creates a small amount of hydrochloric acid in the dish. It won't be a lot, but if you keep your hands submerged in it long enough, you could get a rash. It also creates a small amount of chlorine gas, which is poisonous! Work on a small scale in a large, well-ventilated room. Do not attempt to use this process to make large quantities of metal oxides. Finally, it creates *large* amounts of hydrogen and oxygen gases, which can be explosive if allowed to build up. Again, stick to small scales and work with plenty of ventilation.

4a

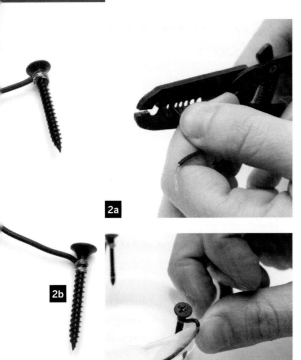

2a

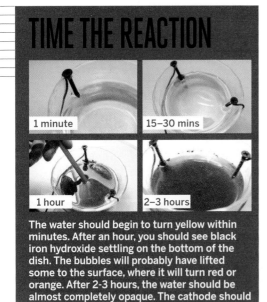

2b

2c

TIME THE REACTION

1 minute

15–30 mins

1 hour

2–3 hours

The water should begin to turn yellow within minutes. After an hour, you should see black iron hydroxide settling on the bottom of the dish. The bubbles will probably have lifted some to the surface, where it will turn red or orange. After 2-3 hours, the water should be almost completely opaque. The cathode should be covered with scale.

Secure the wires to the sides of the dish with tape (**Figure 2c**).

3. **Run the Reaction.**

Plug in the charger. You should see bubbles

coming off the screws. Stir occasionally with a small paintbrush, breaking up any chunks and brushing off the electrodes. Unplug the charger after 3 hours.

(See the sidebar, Time the Reaction)

4. **Separate the Products.**

Remove the electrodes and clean them off. The cathode should be noticeably eroded as shown in **Figure 4a**.

Put on your gloves, take the dish outside, and fill it to the brim with acetone (**Figure 4b**).

Touch the magnet to the bottom of the dish. The black form of iron oxide is slightly magnetic, and this should help separate it at the bottom of the dish (**Figure 4c**).

Leave the dish overnight.

Using Your Pigments · · · · · · ·

The next day, there should be a completely clear layer of water and acetone on top of a thick sludge of red iron hydroxide and black iron oxide. Carefully pour off the excess acetone and water.

NOTE: Used acetone is best disposed of by evaporation: Pour it into a large flat glass or metal receptacle and leave it outside, away from animals and small children, to dry up.

At this point, you have a thin ink you can use as is. Dip your brush, apply it to paper, and the acetone will evaporate quickly without warping the page like water does. The longer you wait, the more acetone will evaporate and

4b

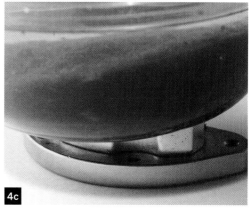

4c

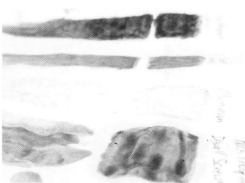

GOING FURTHER

If left uncovered, the acetone will evaporate and leave a dry cake of iron oxide, but you can add more acetone (or some other carrier) to reconstitute it.

If you repeat this process, substituting small copper pipe fittings for the steel screws, you will produce a brilliant orange powder. ◪

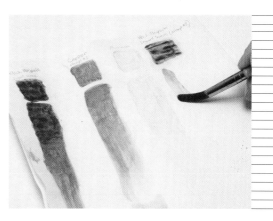

Sean T. McBeth is a hobbyist, software engineer, photographer, and stalwart member of Philadelphia hackerspace Hive76. He can be reached at sean.mcbeth@gmail.com.

The Chronophotonic
Lamp Switcher
Written by *Charles Platt*

⚡ **TIME: 2–3 HOURS**   ⚡ **COST: $20–$30**

MATERIALS

» **Quad comparator IC chip** such as STMicroelectronics LM339N
» **Phototransistor, visible spectrum** RadioShack 55053303
» **Trimmer potentiometers, 500kΩ (2)** such as Bourns 3386F-1-504LF
» **555 timer IC chips (2)** such as Texas Instruments NE555N
» **Ceramic capacitors: 1µF (5), 0.1µF (3)**
» **Relay, SPST or SPDT, 3VDC coil, to switch at least 0.6A at 125VAC** such as Panasonic DS1E-SL2-DC3V
» **Voltage regulators: LM7805 (1), LM7833 (1)**
» **9V alkaline battery** not rechargeable
» **Fuses, 1A 250V AC (2)**
» **Fuse holders (2)**
» **AC adapter, at least 10VDC 100mA**
» **Diodes: 1N4001 (2), 1N4148 (2)**
» **Resistors, ¼W: 1MΩ (2), 220kΩ (1), 100kΩ (2), 10kΩ (5), 3.3kΩ (1), 1kΩ (3)**
» **Digital alarm clock, 3VDC, battery powered** such as Sharp SPC500J

TOOLS

» **Multimeter and test leads**
» **Breadboard**
» **Jumper wires**
» **Soldering iron and solder**
» **Wire cutters and strippers**
» **Stranded wire, 22 or 24 gauge, 2'**
» **Miniature screwdriver**
» **Drill and ⅛" bit**
» **Liquid insulation** aka "liquid electrical tape"

➕ When I go traveling, I like to leave a timer-controlled lamp in my living room to deter burglars. But where I live, the sun sets two hours later in midsummer than midwinter, and I have to remember to adjust the timer with the seasons. Even then, what about stormy nights when the light fades early?

I needed a way to turn on a lamp when outside light levels dim, but off at a fixed time every night when I would normally go to sleep. The "on" function should be controlled by light, the "off" function by a clock. I couldn't buy a gadget to do this, so I built my own.

A Phototransistor and a Comparator

Suppose you wire a phototransistor as in **Figure A**, and point it at the sky. If you recall my key-card door lock from MAKE Volume 33, you'll know that the voltage at point X will gradually fall as the sun sets.

We use a comparator, symbolized in **Figure B**, to convert this gradual change into a sharp, clean signal to turn on a lamp. The + and − signs on the input pins do not mean to apply positive and negative voltage. Rather, both pins accept positive voltage, and if the + input rises higher than the − input, the comparator output flips from low to high — and vice versa.

The LM339 is a cheap, reliable chip containing 4 comparators (**Figure C**), each built around an NPN transistor with "open collector" output. (Some comparators use CMOS transistors with an "open drain" output, but the principle is the same.) To get a

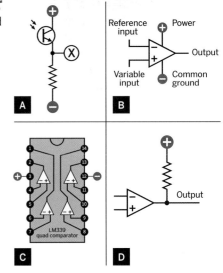

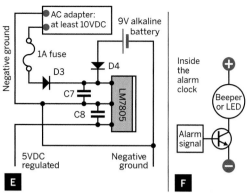

E 5VDC regulated — Negative ground

F Inside the alarm clock — Beeper or LED / Alarm signal

G Beeper or LED — Lamp switching circuit / 1K, 10K, or 100K — Lamp switching circuit

H

positive signal, you attach a "pull-up" resistor, as in **Figure D**.

A "high" output from the comparator is created when the transistor inside it blocks current from the pull-up resistor and diverts it to other components. A "low" output is when the transistor sinks current from the pull-up resistor and takes it from other components.

If we connect the output to a 555 timer wired in monostable mode, the transition will make it emit a single high pulse, which can be amplified by a transistor to activate a latching relay, which can turn on a lamp. (For more information about relays and other components, please see my *Encyclopedia of Electronic Components, Volume 1.*)

The circuit is powered by an AC adapter, with an alkaline battery backup (**Figure E**). In a blackout, the battery will run the circuit for at least 24 hours. In normal use, an AC adapter rated 10VDC or more will prevent the battery from discharging. Be sure it's an alkaline battery, not rechargeable. Don't omit the fuse.

Clocking the Lamp

Now we can switch the lamp on. What about off? The easiest way is to modify a cheap battery-powered digital alarm clock. If you can't find a Sharp SPC500J, just make sure to get one that runs off two 1.5V batteries.

Put in the batteries, open the case, and set the alarm one minute ahead. Press your multimeter's black probe against the negative end of the battery pair, and when the alarm starts, test voltage into and out of the beeper with the red probe. Set your meter to AC volts, since most clocks use a stream of pulses to create an audio signal. Now silence the alarm, and check for DC volts.

In my clock, the beeper has a positive DC voltage on both sides when it is off, because it is wired as in **Figure F**. A transistor triggers the beeper by grounding it, pulling down the voltage on one side. (I have shown the transistor separately, but it's probably built into the chip that runs the clock.) The voltage between the beeper and transistor is normally high; when the alarm goes off, it goes low.

If we tap into the circuit, as in **Figure G**, the voltage transition can activate the 555 timer we use to switch off the relay. Attach an output wire to any point in the alarm wiring where the voltage fluctuates above and below 2V. In some clocks, the beeper may go low-high instead of high-low, but that's OK. The circuit will accept low-high-low or high-low-high, steady or pulsed current, AC or DC, just so long as the high is above, and the low below, 2V. If you can't get that range, substitute a resistor for the beeper (or LED), as shown. Start with 1K and increase as needed.

My clock also illuminates an LED for 5 seconds when the alarm sounds. I decided to use that as my alarm signal instead of the beeper, but the LED didn't create the necessary voltage drop, so I substituted a 100K resistor (**Figure H**).

Details

Figure I shows an overview of the circuit, and **Figure J** the schematic. The clock signal passes through resistor R5 to a second LM339 comparator, triggering a second 555, powering the relay's "off" coil via transistor Q2.

For long-term use, the clock is powered from the switcher circuit with an LM7833 voltage controller. Remove the batteries, solder wires to the battery carrier contacts, and run them

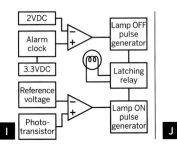

The main circuit, omitting the power supply in Figure E.

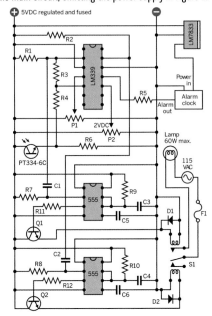

Potentiometers: P1, P2: 500K

Resistors:
R1, R2, R5, R7, R8: 10K
R3: 220K
R4: 100K
R6: 3.3K
R9, R10: 1M
R11, R12: 1K

Capacitors:
C1, C2, C3, C4, C7: 1µF
C5, C6, C8: 0.1µF
(C7 and C8 shown in Figure E)

Fuse: F1: 1A at 250VAC

Transistors: Q1, Q2: 2N2222
or similar

Diodes: D1, D2, D3, D4: 1N4001
(D3 and D4 shown in Figure E)

Relay: S1: Latching dual-coil
3VDC, switching 0.6A minimum
at 115VAC (resistive load)

out through a hole drilled in the case, along with the alarm output wire. Set the reference voltage on the second comparator to 2V using potentiometer P2, and test it with a meter. Now, when the alarm goes off, it should trigger the second 555 timer.

Capacitors C1 and C2 isolate the timers from DC voltage while allowing transitions to pass. Resistors R7 and R8 hold the timer inputs high until a transition occurs. R3 provides positive feedback from the first comparator's output to its input to reinforce the transition and make it fast and clean.

Figure K shows the relay pinouts from above. The transistors powering the coils will cause a slight voltage drop, so I've specified a 3V relay rated for pulses up to 4.5V. Test its outputs with 2 LEDs. Once you've got it working on a breadboard, solder the relay onto a perf board, and check that the joints are secure. Cover them with liquid insulation, and absolutely do not omit the 1A fuse. Use an incandescent bulb no brighter than 60W.

To protect your phototransistor, point it through a window without any direct sun. At sunset, set a very low reference voltage with potentiometer P1, and gradually increase until the lamp turns on. (To test this function, use any bright light that dims gradually.) Now trigger the alarm. The lamp should turn off.

Why Not Use a Microcontroller?

You'd still need 2 transistors to power the relay and an external clock for accurate long-term timekeeping. You wouldn't need a comparator, because the microcontroller could directly process the alarm signal and phototransistor input. You'd have to guess at the software value that corresponds to a "sunset" state, download the program, test it, edit it to adjust the value, etc. Personally, I think twiddling a potentiometer is easier. ◪

Turn lamp ON   To lamp

Turn lamp OFF
K **Relay viewed from above**

NOTE: If you upgrade to house current to switch a lamp, be careful! Never run house current to a breadboard!

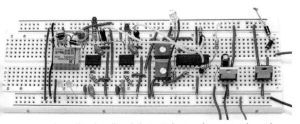

A complete breadboarded circuit showing the orange relay with input transistors and 2 green test LEDs, two 555 timers with red test LEDs, blue trimmer potentiometers, LM339 quad comparator chip, and voltage regulators for 3.3VDC and 5VDC.

Charles Platt is the author of *Make: Electronics*, an introductory guide for all ages, as well as the ongoing *Encyclopedia of Electronic Components* series. He's working on a sequel, *Make: More Electronics*. makershed.com/platt

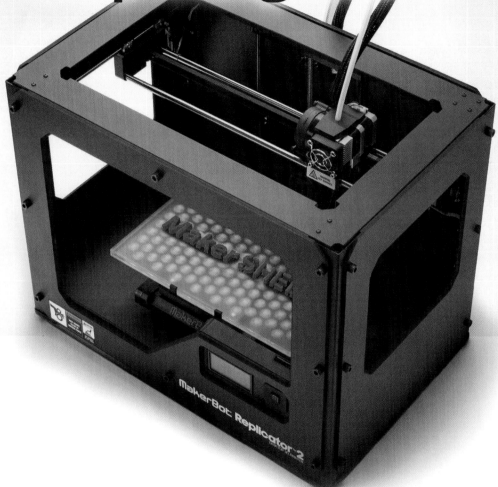

TOOLBOX

BioLite Thermoelectric CampStove
$130 biolitestove.com

» Twig stoves are handy and eco-friendly – you don't haul fossil fuel, just collect sticks wherever you're camped. They're dirt cheap to make, but Brooklyn-based BioLite, a Kickstarter success story, has gone them one better. The BioLite CampStove is a fan-stoked rocket stove with a thermoelectric module that converts heat into electricity. It charges its own battery, and, via USB, charges your cellphone, GPS, headlamp, or other gadgets.

I tested the BioLite and was impressed. Power up the fan and your dry kindling will blaze up easily with a single match. A single charge of hardwood twigs boiled 1 liter of water in just over 4 minutes, comparable to gas canister and liquid-fuel stoves, and burned just long enough to boil 2 liters. Combustion is super efficient; there's a vortex of flame but virtually no smoke, and little ash.

After a third load of twigs, I got the green light for charging my iPhone; it charged tolerably fast (4W max power), but if you want a full charge you'll have to feed the stove continually as a mini-campfire after dinner.

It's heavier (2lbs+) than most backpacking stoves, but the weight cancels out if you're leaving behind liquid fuel, fuel bottles, solar charging gear, and/or spare batteries. Overall it's a great option for car, bike, and boat camping and longer backpacking trips. I plan to stash one in my vehicle, ready for impromptu overnighters or apocalypse bug-out.

–Keith Hammond

Gregory Hayes

Bosch GCM12SD 12" Dual-Bevel Glide Miter Saw

$976 **grainger.com**

This high-end, well-made, surprisingly quiet saw thoroughly exceeded my expectations in terms of agility, compactness, and ease of use. Its 12-inch blade is driven at 3800 RPM on a compound cutting arm that can accommodate straight cuts up to 6.5 inches thick and 12 inches deep. The base has extendable platforms for supporting long stock. It makes very short work of 4-inch PVC.

One of the great features of this saw is that it can move dramatically without taking up much space. The GCM12SD comes with an articulated arm, and can be backed up right against a wall without restricting front-to-back movement. The saw can bevel up to 47° in either direction, and simultaneously miter up to 52° left or 60° right. Even at full bevel and miter stops, it can still handle stock up to 4 inches thick and 12 inches deep.

The only major drawback is the price, which may be prohibitive for weekend projecteers. But anyone who uses a chop saw frequently will probably find that investing in a GCM12SD pays off pretty quickly.

—Nick Parks

Heat Sink Soldering Tweezer

$13 **micromark.com**

The Micro-Mark Heat Sink Soldering Tweezers are an affordable, useful addition to any electronics bench. Each jaw of the tweezers is milled out to grasp small hookup wire perfectly, making it a valuable tool for soldering. The stainless-steel body acts as a large heat-sink, preventing the wire's insulation from melting and keeping heat-shrink from prematurely shrinking. The red wooden grips provide a nice feel, and protect your fingers from the heat.

I also found them useful as all-purpose grippers, especially when dealing with small parts. The milled slots in each jaw are perfect for picking up small bolts, and the jaws' ability to keep things vertical make them great for carefully gluing things together.

—Eric Weinhoffer

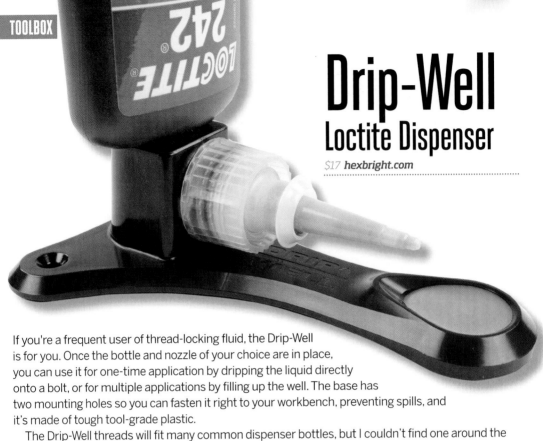

Drip-Well
Loctite Dispenser
$17 *hexbright.com*

If you're a frequent user of thread-locking fluid, the Drip-Well is for you. Once the bottle and nozzle of your choice are in place, you can use it for one-time application by dripping the liquid directly onto a bolt, or for multiple applications by filling up the well. The base has two mounting holes so you can fasten it right to your workbench, preventing spills, and it's made of tough tool-grade plastic.

The Drip-Well threads will fit many common dispenser bottles, but I couldn't find one around the shop that had the correct nozzle length to actually reach the well. Nonetheless, it will get plenty of use with 50ml Loctite bottles, and I can save the long nozzles for possible future use with other liquids.

—*EW*

Tamiya 70097
Twin-Motor Gearbox
$13 *tamiyausa.com*

This inexpensive kit includes two DC motors driving separate axles, providing independent control of speed and direction to left and right. It makes an ideal power plant for a small tracked or two-wheeled robot. The instructions include plenty of diagrams, and

the kit is easily assembled using just a screwdriver and the bundled Allen key. It can be configured in both high-speed (58:1) and high-torque (203:1) gear ratios, and plays well with Tamiya's Track and Wheel (#70100) and Universal Plate (#70157) sets to make a quick and versatile custom robot base. Solder some leads to the motors, add a controller, some sensors, and a power source, and you're well on your way to a sweet little roaming 'bot.

—*Scott W. Vincent*

Liquid Image Explorer 304 Videocam Dive Mask

$100 **liquidimageco.com**

For inspecting your leaky submarine
or snorkeling with the kids, here's a handy toy.
This mask's built-in 8MP camera takes VGA video at up to 30 frames
per second and accepts a microSD card up to 32GB for 20+ hours of video. It's very easy to
operate underwater, with just two big buttons (Still/Video/Off, and Shutter) and a simple status
LED. Peripheral vision is blocked by battery compartments, and it's not the smartest camera —
above water, I found the image contrast varied greatly as my viewpoint bobbed between wave
and sky. And it downloads its files very slowly via USB. But the advantages are obvious; no
separate GoPro to fool with, no helmet mount needed, just grab your video mask and plunge in.
This one's rated to only 15 feet; for deeper dives and better images, consider the 12MP wide-angle
Scuba series, rated to 130 feet, with HD 720P video for $200, or HD 1080P for $300.

—KH

SpareOne Emergency Mobile Phone

$100 **spareone.com**

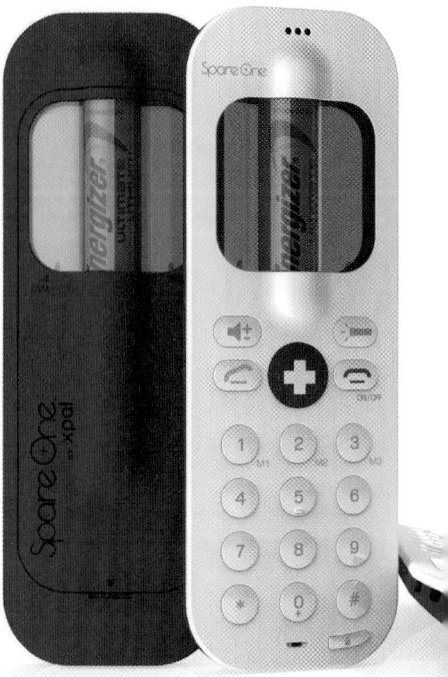

You just blew up your Galaxy attempting to video your rocket launch and you're hurt. Your ROV's adrift and your Nexus slipped overboard. Or, like me, your car's dead and you cooked your iPhone senseless on the hot dashboard.

Call for help with this emergency GSM cellie, with or without your SIM card. It's barebones — no screen! — but it runs on a lithium AA battery that'll last 15 years unused (it provides 10 hours of talk time when in use), survives conditions other phones can't (−22°F to 140°F), and automatically text-replies your friends telling them to phone you instead. And it's unlocked, so you can buy prepaid SIMs too.

An upgrade, the SpareOne Plus, debuted at CES 2013 but hasn't hit stores is said to include a location-tracking service and a waterproof, floatable bag that you can talk through.

—KH

Hauntbox Prop Controller

$179 **hauntbox.net**

The easiest way to automate a haunted house is to bribe your friends to hide behind curtains and pull ropes all night. A slightly more complicated (and considerate) way is the Hauntbox "automation machine." With screw terminals for six digital inputs like break-beam triggers and motion sensors, and six outputs at 5V, 12V, and 24V, you can trigger lights, motors, pneumatics, and all manner of mechanical hauntery. And with a PowerSwitch Tail, you can toggle a 120V AC outlet at will — making the Hauntbox the easiest method of home automation I've found.

The hardware's open source; the dead-simple browser interface lets you program delays, durations, and the optional sound module from any device on your network; and an override tab gives you manual control. Fine-tune your haunt on-the-fly from your phone! Plus you get Arduino compatibility and most of the I/O pins of an Arduino Mega ADK. The barrier between the average Joe and a fully automated haunted house — or just smart house — just got beaten down a notch.

—Sam Freeman

Blocklite 9V LED Flashlight

$10 amazon.com

Fun fact: The 9V, aka "Power Pack 3," battery was originally part of a whole series of Eveready "Power Pack" batteries, most of which featured the signature "snap" terminal connectors. Today the 9V is the only battery in common use that still has these positive-locking connectors. And frankly, they intrigue me. What other clever, tiny devices could we design for batteries to clip onto, instead of into?

Besides the clever clip-on feature, I dig the Blocklite as an exercise in minimalism: it's not much more than a 9V battery snap, six bright white LEDs, and a three-position center-off switch. Switch left to illuminate all six LEDs, right for just the middle two. The bundled Eveready PP3 was still giving usable light after 96 hours of continuous operation on the bright setting. It measures, weighs, and costs just a hair under 1 inch, 10 grams, and 10 bucks, respectively, and totally justifies every unit.

—Sean Michael Ragan

Thermite Kit

$20 skylighter.com

This inexpensive kit has the power to turn sand into glass, burn holes through pavement, and totally demolish old hard drives. The kit consists of powdered aluminum and rust (which you mix in a 1-to-3 ratio) and a box of sparklers (to use as igniters). I used a postal scale to measure the chemicals and combined them in a terra cotta flowerpot, which was a messy process. I put the pot on top of a stack of old hard drives, stuck a sparkler in the powder, and wrapped some hair dryer wire around the sparkler so I could ignite it remotely. I touched the wires to a drill battery and, seconds later, was treated to a spectacular flame and a torrent of molten iron flowing onto the hard drives, rendering them completely useless. Best 20 bucks I ever spent.

—Michael Castor

BOOKS

Books Books Books Books Books Books Books Books Books Books Books Books Books Books Books

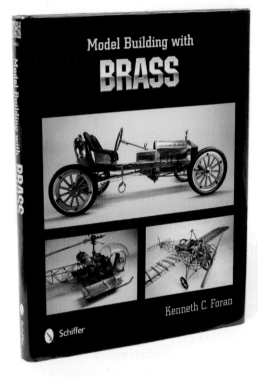

Model Building with Brass
by Kenneth C. Foran
$35 **Schiffer Books**

Brass is a versatile, friendly, fun material for small-scale machine and modeling work. Though brass enjoys wide use in steampunk, antique car, aircraft, marine, and (especially) railroad modeling, no full-length books had been written about it before this title came along in 2012.

Foran is a retired industrial designer and life-long model builder, and the stunning photographs of his work add significant value as inspirational eye candy. The technical content is also rich and fairly deep. After the predictable front matter — details of the material, various common tools, and rudimentary techniques — chapters cover fabricating parts from stock shapes, incorporating other metals like copper, using tabletop machine tools, and chemical processes like cleaning, plating, and etching.

Though the photos are consistently excellent, their relationship to the text is often haphazard, and the text itself clunky. However, the instructional value of this information, together with the inspirational value of the photos, more than make up for these minor annoyances.

—*SMR*

The Unofficial Lego Technic Builder's Guide
by Paweł "Sariel" Kmieć
$24-30 **No Starch Press**

I'm a mechanical engineer, and I wish I had discovered this book as a kid. It covers a wide range of mechanical design concepts — from backlash and efficiency to pneumatic engines and differentials — in the hands-on context of Lego models. A great fit for someone ready to push his or her limits as a builder, or as an introduction to a mechanical design course.

—*Kevin Simon*

The Unofficial Lego Builder's Guide
by Allan Bedford
$20-25 **No Starch Press**

Reading this book reminds me of how Alice-in-Wonderland the world of Lego is. Bedford guides you down the rabbit hole past jumbo blocks, micro-scale figurines, mosaics, and a dozen other styles, tips, and tricks of the trade. Includes both concrete building techniques and a more abstract introduction to the creative process behind world-class Lego model design.

—*KS*

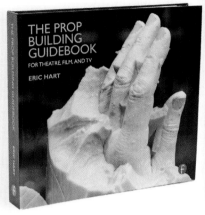

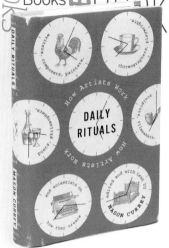

The Prop Building Guidebook: For Theatre, Film, and TV
by Eric Hart

$40 Focal Press

Want to build a haunted house from the ground up? Start with this dense book. And maybe finish with it, too. Yes, it teaches how to make props and scenery, but making a good fake often takes more knowledge than making the real thing.

In a personable, informative style, Hart gives you the full rundown, from basic to advanced techniques across a spectrum of disciplines, supported by excellent tutorial photos as well as theatrical eye-candy shots. Chapters cover safety, shop tools, joining, measuring, materials, sculpture, finishing, and more with rich detail and sparkling clarity. The heavy-duty comb binding is a thoughtful touch. Hart's book makes a valuable shop companion for makers of all types and skill levels.

—Gregory Hayes

Daily Rituals: How Artists Work
by Mason Currey

$24 Alfred A. Knopf

Begun as a blog series by writer and editor Mason Currey, *Daily Rituals* collects hundreds of short glimpses into the creative processes of notable authors, painters, inventors, composers, and others. Punctuated throughout by pithy quotes, the profiles introduce you to those whose work determines their lifestyle (and sometimes the lifestyle of everyone around them), and to those whose lifestyle seems to barely allow for their work. They describe habits dictated by poverty and schedules seemingly overflowing with day-to-day obligations. They reveal personal tricks for courting the muse and coping with writer's block and other demons. Even if they don't inspire you in your own rituals, the book's bite-sized sketches make for a fun, fast, interesting read.

—GH

Getting Started with BeagleBone
by Matt Richardson

$10-15 Maker Media

Many makers love microcontroller platforms like Arduino, but as the complexity of their projects increases, they find themselves needing more power for applications like computer vision. The BeagleBone is an embedded Linux board designed for these makers. It's got built-in networking, plenty of inputs and outputs, and a fast processor to handle demanding tasks. This book introduces both the original BeagleBone and the new BeagleBone Black, and gets you started with projects that take advantage of the boards' enhanced processing power and improved abilities to interface with the outside world.

New Maker Tech

MATTERFORM **3D Scanner**

$583 ***matterform.net***

There's a race on to deliver a desktop scanner that matches the scale and accuracy of today's desktop 3D printers, and Matterform is the current front-runner.

The subject rotates on the scanning platform as an HD camera records the contours of a pair of laser lines moving across its surface. The scanning envelope is a cylinder 7½" in diameter and 9¾" tall. Average scan time is around 3 minutes, and full-color scanning is supported.

Matterform comes with cross-platform software, and can save scans as raw point-cloud data or STL/OBJ meshes. The scanning platform folds up when not in use for easy storage or transport. *—Matt Griffin*

1. Taulman Nylon Printer Filament

$20-$30/lb ***taulman3d.com***

While high-end 3D printers now boast more than 200 materials, typical desktop equipment supports only two or three thermoplastics at a time — usually ABS, PLA, and PVA. As extruder capabilities improve, the list has expanded to include Laywood, HIPS, polycarbonate, and others.

Nylon — a strong, versatile plastic with applications ranging from toothbrush bristles to self-lubricating hardware — has been on the wish list for some time, but has proved challenging to extrude.

Now, Taulman 3D has released a pair of nylon copolymers specifically designed for desktop printers: Taulman 618 ($20 per 1lb spool of 3mm or 1.75mm filament) and the stronger, pricier Taulman 645 ($30). An FDA-approved grade, Taulman 680, is expected to ship this fall.

—MG

2. pcDuino

$60 ***sparkfun.com***

The pcDuino is an Arduino-friendly development board that works like a PC. It can run Ubuntu Linux, Android ICS, or another OS. It sports an HDMI port, two USB ports, an Ethernet jack, and a microSD slot.

The processor is the impressive A10 — a cheap, powerful chip that powers tablets around the world. The board comes preflashed and ready to hack. Just connect TV, keyboard, and mouse, and it's ready to use like any PC.

The board's "duino" personality manifests as four sets of male pin headers along one side, matching the functions of Arduino's familiar analog and digital pins. You can plug sensors and motors into these ports and enter code using the built-in IDE, which compiles automatically. Note that the header footprint is not the same as an Arduino, so your shields will need adapters to connect.

—John Baichtal

3. Spikenzie Labs Calculator Kit

$45 ***makershed.com***

It may not look like an Arduino, but this kit has an ATmega328-PU, the same IC used in the Uno, for a brain. It looks great sitting on your desk, and you can reprogram it to add functions using the Arduino IDE. I'm modding mine with a stopwatch function featuring a flashy animation.

—Marc de Vinck

4. Bricktronics Shield Kit

$35 ***wayneandlayne.com***

This cool kit lets you connect two Lego NXT motors and four NXT sensors to your Arduino. Goes great with our book *Make: Lego and Arduino Projects*, which covers all you'd ever want to know about creating robots with your Arduino and Lego NXT.

—MV

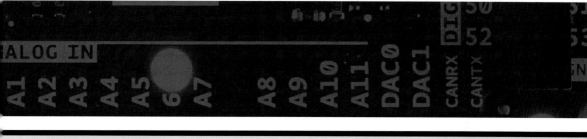

5. PiLarian

$39 *pilarian.com*

This stylish 3D-printed Raspberry Pi enclosure was inspired by the mineral skeletons of micro-organisms called radiolarians. According to the designers, the pattern of holes in the undulating surface conserves materials without sacrificing strength. The piLarian comes in six colors and is printed-to-order by Shapeways.

—*Matt Richardson*

6. PiFace Digital

$30 *pifacedigital.wordpress.com*

This Pi add-on board has eight digital inputs, four buttons, and eight digital outs (with LED indicators). Two outputs control onboard relays for switching high-voltage devices. All terminals have screw connections for secure wiring.

The software includes Scratch, Python, and C examples, plus an on-screen interface for board control and input readout.

—*MR*

7. RaZberry

$69 *razberry.z-wave.me*

Want to use the Raspberry Pi for home automation? Check out the RaZberry, a Z-Wave daughterboard for the Raspberry Pi. This small circuit board connects to the expansion header on the Pi and enables wireless device control using the Z-Wave protocol, an international wireless communication standard for home automation.

The software is easy to install on the standard Raspbian distribution of Linux for Raspberry Pi, and includes examples to help you roll your own interface for controlling your home's lighting, thermostat, locks, security sensors, and appliances. Of course, you'll also need to invest in Z-Wave hardware around your home, but there's a wide selection of devices available at zwaveproducts.com and many hardware stores.

—*MR*

8. 3Doodler

$99 *the3doodler.com*

This Kickstarter favorite is essentially the extruder head from a fused-filament 3D printer cut loose from its robot frame and mounted in a pen body for free-hand use. The operator controls the temperature and flow of plastic to trace out spindly lines or build layers by hand. Many early 3D printer hackers experimented with so-called "freeprinting," but the 3Doodler team of seasoned toy and interactive project designers Peter Dilworth and Max Bogue took the idea one step further.

The 3Doodler can extrude 3mm PLA or ABS filament from bundled color packs or widely available printer feedstocks. Some of the unique applications explored by early testers — like embellishing existing 3D-printed and stock plastic parts — seem promising, though claims that the 3Doodler can also be used for simple home repairs leave me a touch skeptical.

—*MG*

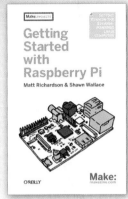
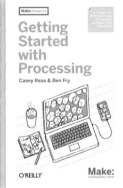
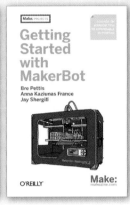

Make: Marketplace

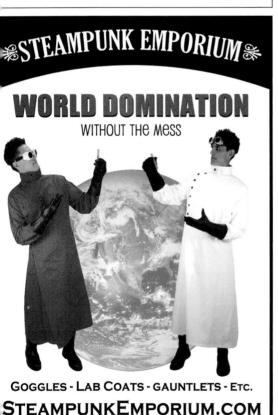

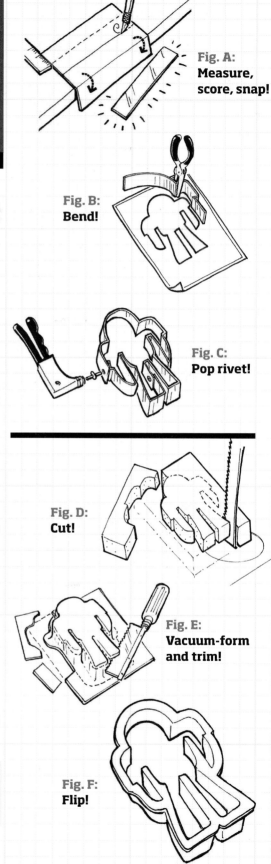

Invented & drawn by **Bob Knetzger**

TOY INVENTOR'S NOTEBOOK

Custom Cookie Cutters

Make your own custom shaped cookie cutters two ways!

Draw your cookie shape outline to use as a same-size pattern. Keep it simple: small details are OK, but bold shapes work best. For metal cookie cutters, use a 0.015" aluminum sheet, available at hardware stores. Measure 1"-wide strips, then score firmly with an X-Acto knife. Bend the score against the edge of a table, then bend it back and forth until it splits. Repeat to make as many strips as needed (**Figure A**). Sand the edges to remove any sharp burrs.

Use needlenose pliers to hold the strip and carefully make bends to match your pattern (**Figure B**). Use a pencil or dowel inside to form tight corners. The soft aluminum bends easily and holds its shape. Keep your bends perpendicular to the strip! To make longer strips or join the ends, drill matching ⅛" holes. Use pop rivets with backing washers to fasten strips together (**Figure C**). Done!

To make plastic cookie cutters, first cut your shape out of wood (**Figure D**), then vacuum-form plastic over it (*see MAKE Volume 11, "Kitchen Floor Vacuum Former,"* makezine.com/go/vacuumformer). Cut out around the shape, leaving a generous rim, then cut or grind off the top (**Figure E, F**). ◪

➕ See pics of cute Bug Boys character-shaped cookie cutters in action at makezine.com/35.

Fig. A:
Measure, score, snap!

Fig. B:
Bend!

Fig. C:
Pop rivet!

Fig. D:
Cut!

Fig. E:
Vacuum-form and trim!

Fig. F:
Flip!